# THE MOTHER EARTH NEWS® ALCOHOL FUEL HANDBOOK



**edited by Michael R. Kerley
and the staff of THE MOTHER EARTH NEWS®
illustrated by Charlie Lawing
and the staff of THE MOTHER EARTH NEWS®**

**THE MOTHER EARTH NEWS®, Inc.**
Hendersonville, North Carolina

Published by THE MOTHER EARTH NEWS®, Inc.
P.O. Box 70, Hendersonville, North Carolina 28791

Printed in the United States

First paperbound printing

ISBN 0-93842-00-1

Library of Congress Catalog Number: 80-84384

# CONTENTS

# 1 INTRODUCTION

In 1949, S.C. Prescott and C.G. Dunn had this to say about the fuel situation: "The subject of ethyl alcohol production by fermentation has assumed new interest on account of attempts to find substitutes for gasoline. Blends of alcohol with gasoline, especially a blend containing 10% ethyl alcohol, may be used satisfactorily in the present internal combustion type of motor. Present-day demands for motor fuels are great, but the sources of petroleum are limited." (*Industrial Microbiology*, McGraw-Hill Book Co., Inc.)

So much for today's "energy crisis". The simple fact is that there's never been a time in the history of the world when there *wasn't* such a crisis. Energy—whether it comes from wood, coal, gas, petroleum, or the movement of water and wind—has always been in short supply.

The United States—with its huge reserves of coal and initially large fields of gas and oil—has long been one of the fortunate societies that had energy in plenty. And that's *still* the case, but—because of the depletion of our own oil and gas fields and our own dependence on such fuels—we have turned to sources of supply outside our borders (and beyond our control) to satisfy our energy needs. Some of those sources are owned by societies actively hostile to ours and others by lands which simply put their own interests ahead of those of *our* nation.

Our whole "imbalance of energy trade" would seem to be part of some inevitable historical process were it not for one simple fact: The frantic international dealings and counterdealings brought about by our current fuel shortage are totally unnecessary. The United States possesses a sufficient amount of energy . . . but it occurs in a variety of forms which we haven't fully exploited (Fig. 1-1).

The dollar is down . . . and gold—an essentially worthless, nonproductive substance—has risen in value, simply because the United States persists in *buying* something it doesn't really need at prices that, in the long run, it can't possibly afford. If the U.S. were to invest as much money (currently around $50 billion per year) in developing *new* sources of energy as it does in lining national treasuries and corporate pockets here and abroad with dollars spent on petroleum products, the "energy crisis" would have about as much relevance to our children as the Depression of the 1930's has had to the generation of the post–World War II "Baby Boom".

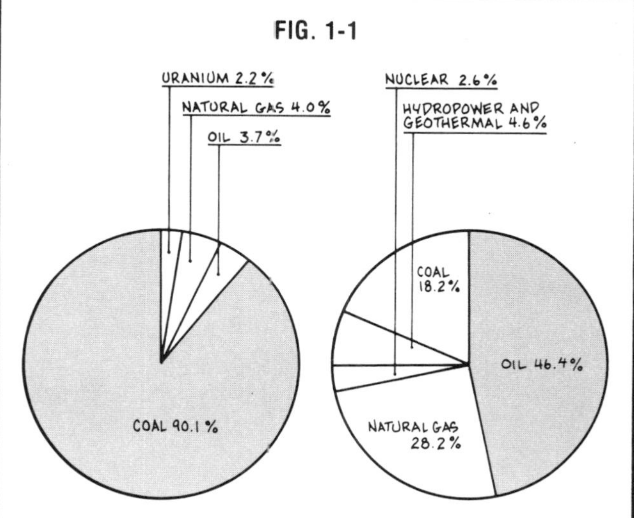


FIG. 1-1

PROVEN FUEL RESERVES VS. 1975 CONSUMPTION PATTERN

Source: U.S. Department of Interior, Bureau of Mines

And where would this abundant energy come from? Much of it can be produced in the fields of America . . . and we *don't* mean the coal fields!

The fuel of the future is *alcohol* . . . which is called ethyl alcohol (or ethanol) when produced from the starches and sugars in vegetable matter . . . and methyl alcohol (or methanol) when obtained by the distillation of wood or—synthetically—from carbon monoxide and hydrogen. As the greatest agricultural nation in the world, and one of the planet's largest wood producers, it's absurd for the United States to remain dependent upon a rapidly diminishing resource when we can supply much of our nation's energy with farm-produced, *renewable* fuel.

But the government in general (with a few very valuable exceptions) seems to be in no hurry to investigate alcohol fuels . . . partially because of the simple fact that the multinational oil companies are extremely hostile to the concept. And usually, when the oil industry is hostile to a program or idea, so will a significant portion of the U.S. government be.

Still, the idea of alcohol fuels is gaining ground, because it appeals to the *people* . . . and does so for the very same reasons that it scares the oil industry: The production of alcohol fuels—particularly ethanol—is an inexpensive, relatively simple technology which can best be handled at small facilities scattered around the nation, rather than at large, centrally located, multimillion-dollar refineries. For this reason, alcohol fuel production may well be impossible to monopolize.

## A SHORT HISTORY OF ALCOHOL FUELS

The idea of using our croplands to produce fuel isn't new. In fact, Henry Ford was one of the first to propose it. As a man who made his fortune through the creation of a new system (assembly line production) and the product the new system made possible (the inexpensive motorcar), Ford distrusted the oil barons—as exemplified by John D. Rockefeller—who had made their money through price-fixing and monopoly. His dream was that the American motorcar would run on grain-produced alcohol —ethanol—manufactured at relatively small distilling operations scattered about the countryside. Then, he said, the country would see "as great a development in farming as we have had in the past 20 years in manufacturing".

But Ford never wielded the national political power that the oil industry has always exercised, and neither did the various farmers' groups that pushed the idea of "Agrol" (a gasoline-ethanol blend) during the Depression of the thirties: a time when farmers were watching their crops rot unbought in the fields (while their children went without shoes for lack of hard cash), or even losing their farms for want of enough money to pay the mortgage. Oil industry campaigns convinced the nation that alcohol fuels would be impractical and too expensive, and that such fuels were the products of an attempt to raid the pockets of the American public . . . a familiar argument!

## THE POTENTIAL AND USES OF ALCOHOL FUELS

The only reason that the idea of alcohol fuels wasn't dead and buried long ago is that such forms of liquid energy *do* offer viable, attainable solutions to many of the fuel problems we continue to have. They are mechanically efficient, available, and—most important—a renewable resource. Fuels made from crude oil—on the other hand—*do* meet the first condition, but are in a "crisis" state on the second, and fail the third completely.

As you read this book you will learn more about the potential of alcohol fuels, but let's establish a few basic facts right away:

[1] The use of grain to produce ethanol is *not* an "immoral" act in a starving world. When the solids which are the by-products of the distilling process are recovered, they contain *100% of the protein value of the original amount of produce* concentrated in two-thirds of the bulk. This material can be used as high-quality livestock feed (approximately 80% of the corn produced in the U.S. is fed to animals), or it can be processed for human consumption. In addition, low-quality, distressed, moldy, and sprouted grains can *also* be used in distillation, and their protein will actually be *recovered* in similar quantities. Such grains are generally considered unfit for human *or* animal consumption. Therefore, distillation can *restore* their value!

[2] The use of crops in alcohol production would effectively end the grain surplus conditions which lead to price supports and farmer bankruptcies.

[3] The relatively small, geographically decentralized distilleries which will best serve this kind

*American croplands provide the raw material for alcohol fuel production. The distillation process will return 100% of the grain's protein value and can utilize even low-quality produce.*

of production would provide hundreds of thousands of new, local jobs.

[4] Alcohol fuels are less polluting than our other available liquid power sources.

[5] Engines which burn alcohol tend to last longer and require less maintenance.

[6] U.S. production of its own liquid fuels would strengthen the national economy *and* make it less vulnerable to foreign political pressure.

All of the above advantages are not, of course, meant to imply that there are no problems presented by the production of fuel ethanol. Small technical modifications are necessary before pure alcohol can be used as a fuel in automobile engines (these modifications are dealt with in Chapter 7). Difficulties are also implicit in the conversion of *any* relatively small industry (such as that currently producing alcohol for beverage and industrial purposes) to a large one (which would be needed to supply a significant part of the nation's liquid fuel requirements). There is also the problem of desperate opposition from entrenched economic interests. This last barrier is likely to be the hardest to surmount . . . if, in fact, it is overcome at all, it will probably be because *alcohol fuels are so easy to produce that almost anyone can make them*!

## ABOUT THIS BOOK

The purpose of this book is to let you in on the "how to do it" side of the home production of ethanol (the technology necessary for methanol production isn't practical for the small producer yet . . . nor is the "other" alcohol quite as efficient a fuel as ethanol is). You will find in this book a practical "nuts and bolts" approach to the technique and technology of producing your own liquid fuel. By the time you've finished reading, you should have a concrete, realistic idea of the resources you'll need and the techniques you'll use in order to make enough "liquid energy" to supplement or satisfy your own needs . . . or even to produce enough ethanol for sale to friends and neighbors (or to larger *commercial* distilleries for conversion into 100%—anhydrous—alcohol).

As the price of crude oil–based fuels continues to rise—while their availability is a constant question with an unsure answer—the need for finding alternate liquid energy sources to fuel home heating units and internal combustion engines is becoming increasingly obvious.

# 2 BACKGROUND IN ETHANOL PRODUCTION

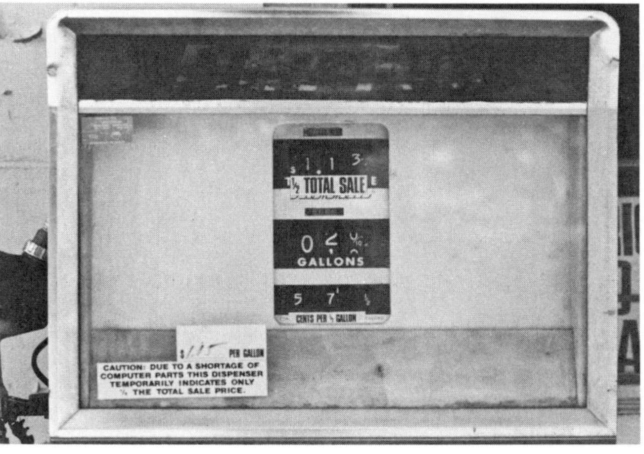


*This gasoline pump—which registers only one-half of the total sale price—is a compelling reminder of the United States' fuel dependence on petroleum supplies in foreign countries.*

## THE ARGUMENT FOR ALCOHOL FUELS

Before we get into the mechanics of producing alcohol at home, we will say a few words about what makes ethanol a viable liquid fuel for the 1980's (when it *wasn't* considered so earlier), and about just what ethanol *is* and where it comes from.

As we pointed out in the introduction to this book, the idea of using alcohol fuels—and using ethanol in particular—isn't a new one. Henry Ford thought it was a great idea, and a whole lot of farmers living in the midwestern corn and wheat states during the years of the Great Depression thought so too. But inexpensive gasoline, the relatively minor mechanical problems encountered in running internal combustion engines on ethanol, and the argument that it took more energy to produce ethyl alcohol than could be realized from it (both the fuel needed to produce the product and the *food* energy in the raw material used were considered) pretty effectively sank ethanol as a commercially viable fuel. And government regulations on the owning and operating of stills completed the job.

Of course, many feel that it was the large oil companies and their influence that led to the public dismissal of ethanol as a viable fuel alternative. During the Depression, midwestern farmers—stuck with huge grain surpluses and no cash—pushed hard for the acceptance of alcohol fuel production, and many observers and activists during that time charged that only the concentrated campaign against such an industry by Big Oil prevented its emergence. Even today, oil industry spokesmen are hostile to the idea of producing and using ethanol as a liquid fuel, claiming that alcohol is wasteful of energy and mechanically inefficient (see Chapter 7 for a discussion of this latter point).

The argument that more energy is spent in producing alcohol from grain than is contained in the final fuel is often true, but the statement misses an important point. We have a surplus of the grain needed to produce alcohol and a shortage of liquid fuel. Or as Dr. George Tsao of Purdue University's Laboratory of Renewable Resources Engineering has pointed out:

"We have a shortage in this country of *liquid fuel*, and we spend nearly $50 billion per year to import it. . . . *Liquid fuel* . . . is needed immediately" (our italics).

*And ethanol is a liquid fuel.* Instead of spending

*This vacuum-operated distillery has the capacity to recover carbon dioxide produced during the distillation process. All of the tanks—as well as the still's vat—are made of fiberglass.*

our money to import the fuel, we can produce it ourselves.

Furthermore, alcohol production need *not* result in expensive fuel. As evidenced by the phenomenal growth of (and extensive big-capital investment in) the alcohol fuel business of late, even a large distilling operation—one which burns fossil fuels in order to produce ethanol—can realize a substantial after-tax profit *if* the by-products of the production process are conserved and utilized. We'd like to emphasize that—by recovering *all* of the by-products and residues of the distillation process (i.e., ethanol, distiller's dried grains and solubles, and carbon dioxide)—alcohol production can turn our "energy crisis" into an important economic (as well as environmental) boon for our entire nation. (Fig. 2-1 is an illustration of the inputs and outputs of the process, while Chart 2-1 is a forecast balance sheet produced by a Nebraska study team.)

Even more to the point for readers of this book, however, the *individual* small farmer or home distiller can produce ethanol more efficiently than can a large operation, because—due to his or her use of relatively small production runs and the availability of alternate waste fuels—he or she can burn such waste products as corn stalks and cobs, wheat straw, and so on . . . as well as wood. Another very important point is that the ingenious distiller who is producing ethanol in fairly large quantities (and thus generating large amounts of heat to run his or her boilers) should find it no difficult task to recirculate the waste (exhaust) heat from the process to heat a home or outbuildings . . . or to *preheat* the cooking water for the *next* batch of alcohol.

And aside from the thermal issue, a number of other advantages are inherent in the production and use of ethanol as a fuel, both to the individual small distiller *and* to the nation. Here are some of the most important:

**Grain Surpluses**—The ever-increasing efficiency of the American farmer has led to low grain prices (which are not necessarily reflected at your corner grocery, because of correspondingly big processing and packaging costs . . . i.e., whether you buy such highly processed and packaged "foods" as TV dinners or not, you *pay* for them), price supports, and much acreage taken out of production. Growing feed grains for fuel-making would give the farmer a high-efficiency cash crop (in terms of yield) with an assured market.

A PRACTICAL FARM ENERGY FLOW CHART
FARM EQUIPMENT TILLS SOIL
FUEL FOR FARM EQUIPMENT
HOME HEATING FUEL
ORGANIC FERTILIZER
FEED CORN AND POTATOES
ALCOHOL PLANT
WATER IS RECYCLED
WASTE* WOOD
CORNCOBS WASTE OIL OLD TIRES*
METHANE PLANT*
DDGS
LIVESTOCK
MANURE
$CO_2$
WASTE HEAT
* POSSIBLE STILL HEAT SOURCES
FIG. 2-1

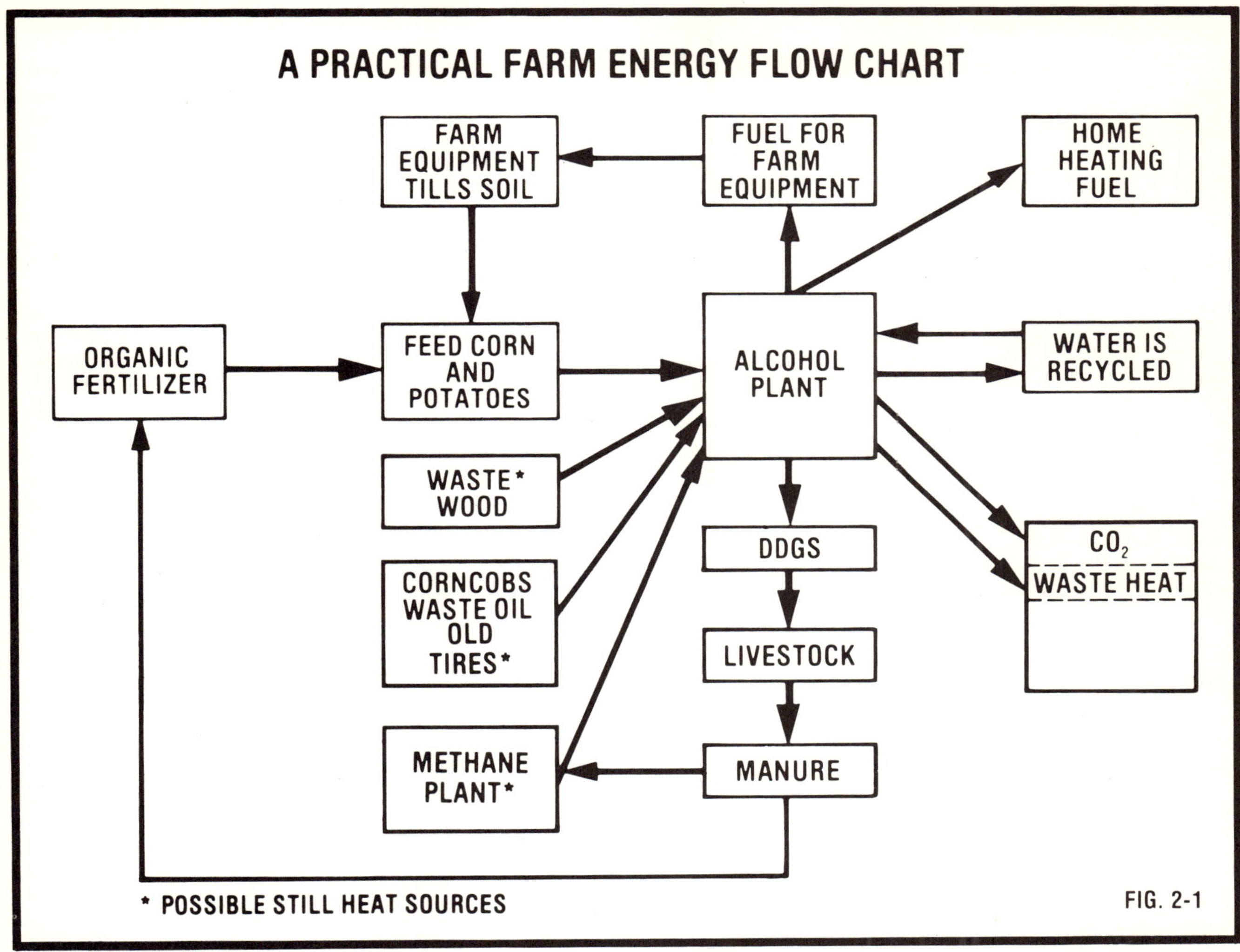

**MATERIAL BALANCE AND ECONOMICS (For a Hypothetical Plant Producing 20 Million Gallons/Year of 200-Proof Ethanol)**

| | Conventional Plant | With Protein Recovery |
|---|---|---|
| ***Investment:*** | | |
| Plant Investment | $23,000,000 | $30,600,000 |
| Working Capital | 4,000,000 | 4,400,000 |
| Total Investment | $27,000,000 | $35,000,000 |
| ***Income:*** | | |
| Ethanol, 20 million gal./yr. @ $1.13/ gal. | $22,600,000/yr. | $22,600,000/yr. |
| Distiller's Dried Grains | | |
| 241 tons/day @ $120/ton | 10,560,000 | |
| 199 tons/day @ $95/ton | | 6,900,000 |
| Carbon Dioxide 199 tons/day @ $2/ton | 150,000 | 150,000 |
| Protein Concentrate 74,000 lb./day @ 42¢/lb. | | 11,340,000 |
| Total Income | $33,310,000 | $40,990,000 |
| ***Expenses:*** | | |
| Milo, 24,620 bu./day @ $3.50/cwt. | $17,610,000 | $17,610,000 |
| Conversion Cost | | |
| @ 30¢/gal. ethanol | 6,000,000 | 6,000,000 |
| @ 14¢/lb. protein | | 3,780,000 |
| Total Expenses | $23,610,000 | $27,390,000 |
| ***Depreciation:*** (10% straight line) | $ 2,300,000 | $ 3,060,000 |
| ***Taxes:*** (50%) | $ 3,700,000 | $ 5,270,000 |
| ***Net Cash Flow:*** | $ 6,000,000 | $ 8,330,000 |
| As Percent of Investment | 22.2% | 23.8% |

**CHART 2-1**

**"Distressed Grains" Recovery**—A large percentage (up to 20%) of the grain grown in the U.S. *cannot* be marketed as human food *or* as animal feed, because it is "distressed" . . . that is, cracked or moldy and therefore determined unfit for consumption. However, such grain *can* be used for ethanol production in the same way that "good" crops can . . . and the residual solids can be recovered and *utilized*, since the mashing and fermentation process serves to purify them.

**Distiller's Dried Grains and Solubles (DDGS)**—The solids left over after the fermentation process (i.e., the matter that is separated out before or after distilling) are made up of a high-protein concentrate which can be used directly as animal feed (or processed into a food supplement that is palatable to humans). DDGS is up to 34% protein (see Chart 2-2) . . . and approximately one-third bushel of this high-grade "builder-upper" can be recovered from each bushel of corn which is used for the manufacture of ethanol. Tests made in 1955 showed that, with hay and corn rations, 474 pounds of beef per 100 bushels of corn were produced . . . while—when cattle were fed hay, corn, *and* DDGS rations—*535 pounds* of beef were produced per *80 bushels* of corn *added to* the DDGS from 20 bushels of corn. That's a net gain of 61 pounds (12.9%) per hundred bushels of corn . . . without even considering the "bonus" ethanol produced from the 20 DDGS bushels!

Commercially, DDGS is dried and sold to farmers by distilleries as a high-efficiency feed supplement for their cattle . . . and it's easy to see why! Better yet, the farmer who manufactures his or her own ethanol, and so produces feed from spent mash, doesn't even need to expend energy in drying the animal feed . . . since he or she has no storage or transportation problems.

**Recovery of Waste Materials**—The wastes produced in the course of grain processing (such as corn stalks and cobs, wheat straw, chaff, etc.) can be used to fire boilers in the production of ethanol.

**A Renewable Resource**—The simple fact is that ethanol—like the materials that go into its production—is a renewable resource, while fossil fuels are not. It may be comforting (or at least satisfying) to shout "conspiracy" and point fingers at the oil companies, the government, and the OPEC nations as gasoline prices rise and the supply is restricted, but fossil fuels—gasoline, heating oil, and even coal—are finite, and thus *bound* to get increasingly scarce *and* increasingly more expensive. Both the individual and the nation are fuel-using and energy-dependent. The use of renewable resources for the generation of liquid fuels could help to solve many grave political, economic, and environmental problems.

## MAKING ETHANOL . . . A LOOK AT THE BASIC CHEMICAL PROCESS

Ethyl alcohol (ethanol) is an organic compound, the molecular formula of which is $C_2H_5OH$ (sometimes written $CH_3CH_2OH$). It can be thought of as a close cousin to methyl alcohol ($CH_3OH$) and ethane ($C_2H_6$). Ethyl alcohol—for our purposes, at least—is produced by the fermentation and distillation of natural sugars . . . whereas methyl alcohol (methanol) is produced—by high-technology methods—from wood, coal, cellulosic wastes, etc., and ethane is a by-product of distilling gasoline (many oil firms produce ethanol *from* ethane).

Ethyl alcohol results from the action of the yeast plant upon simple sugars such as glucose. Another by-product is carbon dioxide ($CO_2$). Chemically, the process can be rendered like this:

$$\underset{\text{(glucose)}}{C_6H_{12}O_6} \longrightarrow \underset{\text{(ethyl alcohol)}}{2C_2H_5OH} + \underset{\text{(carbon dioxide)}}{2CO_2}$$

Glucose (and other simple sugars such as fructose and maltose), in turn, can be found in mediums such as fruit and sugar cane, or can be produced through the action of chemical biocatalysts (known as enzymes) upon *starches*.

Though both ethanol and methanol *can* be produced from a wide variety of biological source materials, and though both are excellent liquid fuels for use in boilers and internal combustion engines, methanol—at present—can be economically produced only in large, commercial-scale operations. This is because its economics are presently those of quantity production, involving complex equipment and techniques which are beyond the resources of the small producer. Because of the economic disadvantages of methanol production, this book will devote itself entirely to the home production of ethanol.

### PROXIMATE TOTAL DIGESTIVE NUTRIENTS OF DISTILLER'S FEEDS*

| | CORN | | | MILO |
|---|---|---|---|---|
| | Dist. Dried Grains | Dist. Dried Solubles | Dist. Dried Grains With Solubles | Dist. Dried Grains |
| % Moisture | 7.5 | 4.5 | 9.0 | 10.0 |
| % Protein | 27.0 | 28.5 | 27.0 | 34.0 |
| % Fat | 7.6 | 9.0 | 8.0 | 8.0 |
| % Fiber | 12.8 | 4.0 | 8.5 | 13.0 |
| % Ash | 2.0 | 7.0 | 4.5 | 4.0 |

Source: Distiller's Feed Research Council, Cincinnati, Ohio.

NOTE: The remaining percentages are composed of bulk (cellulose and other nitrogen-free extracts).

*Distiller's feeds should not be fed as a whole or complete ration. They should constitute no more than 30% (wet or dry) of the total when mixed with other feeds. Since each farmer has particular feed stocks, he or she should consult with the county livestock agent to determine the best formulation for his or her specific livestock needs.

CHART 2-2

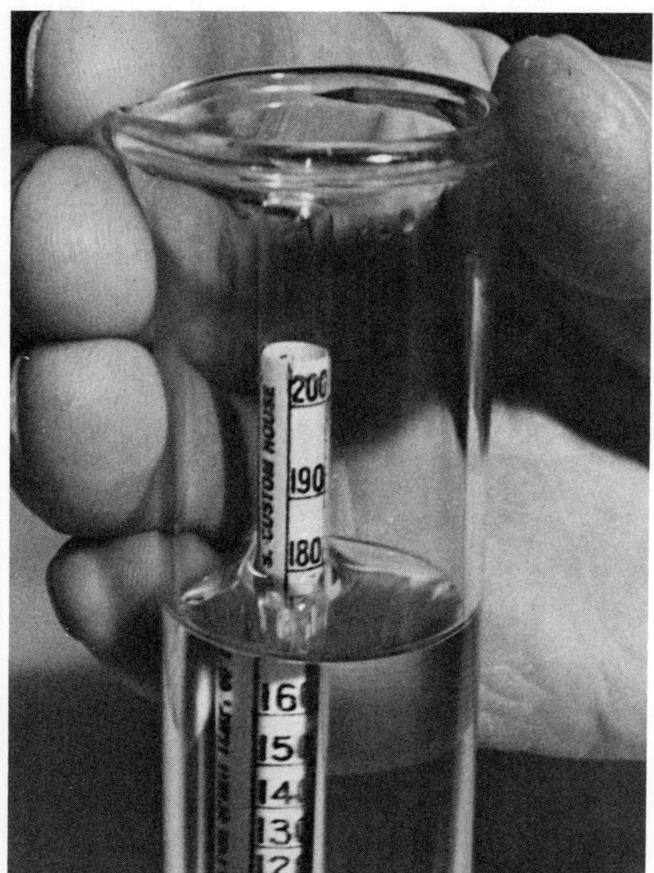


*The hydrometer is an indispensable tool in home-scale ethanol production. The instrument enables the fuel-maker to measure the proof of the alcohol after it has been distilled.*

## A FEW WORDS ABOUT "PROOF" AND OTHER UNITS OF MEASUREMENT

The units of measurement commonly used in describing ethanol production are the terms "proof", "proof gallon", "tax gallon", and "wine gallon". With the exception of "proof", such words are of specific relevance to ethanol produced for beverage purposes. They are defined as follows:

**Proof**—a way of describing the amount of alcohol in a solution . . . "proof" is simply twice the percent, by volume. Thus a solution which contains 95% alcohol is 190 proof.

**Proof Gallon**—a standard U.S. gallon that contains 100-proof (50% by volume) ethyl alcohol.

**Tax Gallon**—equivalent to a proof gallon for a solution containing a concentration of ethanol of 100 proof or more, or equivalent to a wine gallon for a solution containing a concentration of less than 100 proof. An appropriate tax rate is assigned to each.

**Wine Gallon**—a gallon containing 40-proof (20%) alcohol.

The measurement which is most important to the distiller of ethanol fuel—that of proof—is easy to determine without chemical analysis. The specific gravity of ethanol is 0.794. (The specific gravity is the ratio of the density of a substance to that of any other substance . . . water being the standard substance used in comparing liquids.) The specific gravity of water is 1.000. Thus, ethanol is lighter than water, and a solution which combines the two will have a lower density than will pure water.

Hydrometers are tools used to measure the specific gravity of a solution . . . and thus its proof, or ethanol content. You can use specialized 0–200 proof hydrometers for this purpose. Or if you choose to use hydrometers which are graduated only in terms of specific gravity, your instrument must be used in concert with a specific gravity table such as the one reproduced in Chart 2-3.

Because the charted figures are accurate only for readings taken at 68°F, and because the calculations for readings taken at other temperatures are rather complex, we suggest you allow your alcohol to cool to 68°F, if possible. However, here is a step-by-step method for determining actual proof at temperatures other than 68°F.

For the sake of providing an example, assume that the temperature is 80°F.

[1] If the temperature of the solution is 80°F and you find that its specific gravity measures out to 0.8924, you can consult the specific gravity table to find that its percentage of alcohol is *about* 69%, i.e., the solution is about 31% water. Then consult Chart 2-4.

[2] On the table, find the correction factor for the temperature of your solution. Since our hypothetical sample happens to be at 80°F, the correction factor in this case is 1.0069. Multiply that figure by the original specific gravity (0.8924) to reach the *corrected figure* (1.0069 X 0.8924 = a total of 0.8986).

[3] Now return to the specific gravity chart (Chart 2-3) and seek out a figure that's close to our temperature-corrected 0.8986 number. Our new specific gravity falls between 66 and 67% alcohol, so we can say that the actual proof of the distillate is approximately 133. (Note, too, that if the temperature had been less than 68°F, instead of higher, the corrected proof of the alcohol/water solution would actually have been increased instead of decreased.)

If you would prefer using a simpler method, you can order the *Gauging Manual* from the U.S. Printing Office and use it with your hydrometer (see the recommended reading list at the back of this book).

## SOURCE MATERIALS FOR ETHANOL

Because of its relative efficiency as a source of ethanol—as well as its abundance and reasonably low cost—corn is the raw material which is most often associated with ethanol production. However, a number of biological products can be used for the production of ethanol. Chart 2-5 illustrates the relative productivity of various source materials in yielding 99.5% pure ethanol per ton of input. Chart 2-6 illustrates the relative productivity per acre.

It is very unlikely, however, that the individual distiller will realize precisely such outputs, for two reasons. First, 199-proof alcohol is considerably more difficult to distill . . . and of course yields less volume than does 190 proof. Second, the figures appearing on these charts are based on the experience of large commercial distilleries using expensive technology that's not practical for the small distiller. Nevertheless, the general efficiency relationships between source materials which the tables reveal are consistent.

Of course, relative yield is not the *only* factor to

### PERCENT ALCOHOL/SPECIFIC GRAVITY

| Alcohol Percent | Specific Gravity | Alcohol Percent | Specific Gravity | Alcohol Percent | Specific Gravity | Alcohol Percent | Specific Gravity |
|---|---|---|---|---|---|---|---|
| 0 | 1.0000 | | | | | | |
| 1 | .9985 | 26 | .9698 | 51 | .9323 | 76 | .8746 |
| 2 | .9970 | 27 | .9687 | 52 | .9303 | 77 | .8720 |
| 3 | .9956 | 28 | .9676 | 53 | .9283 | 78 | .8693 |
| 4 | .9942 | 29 | .9665 | 54 | .9263 | 79 | .8666 |
| 5 | .9928 | 30 | .9653 | 55 | .9242 | 80 | .8638 |
| 6 | .9915 | 31 | .9642 | 56 | .9221 | 81 | .8610 |
| 7 | .9902 | 32 | .9630 | 57 | .9200 | 82 | .8553 |
| 8 | .9890 | 33 | .9617 | 58 | .9178 | 83 | .8528 |
| 9 | .9878 | 34 | .9604 | 59 | .9156 | 84 | .8524 |
| 10 | .9866 | 35 | .9591 | 60 | .9134 | 85 | .8494 |
| 11 | .9854 | 36 | .9577 | 61 | .9112 | 86 | .8464 |
| 12 | .9843 | 37 | .9563 | 62 | .9089 | 87 | .8434 |
| 13 | .9832 | 38 | .9548 | 63 | .9066 | 88 | .8402 |
| 14 | .9821 | 39 | .9533 | 64 | .9043 | 89 | .8371 |
| 15 | .9810 | 40 | .9518 | 65 | .9020 | 90 | .8338 |
| 16 | .9800 | 41 | .9502 | 66 | .8997 | 91 | .8305 |
| 17 | .9789 | 42 | .9486 | 67 | .8973 | 92 | .8270 |
| 18 | .9779 | 43 | .9469 | 68 | .8949 | 93 | .8235 |
| 19 | .9769 | 44 | .9452 | 69 | .8924 | 94 | .8198 |
| 20 | .9760 | 45 | .9435 | 70 | .8900 | 95 | .8160 |
| 21 | .9750 | 46 | .9417 | 71 | .8875 | 96 | .8121 |
| 22 | .9740 | 47 | .9399 | 72 | .8850 | 97 | .8079 |
| 23 | .9729 | 48 | .9381 | 73 | .8824 | 98 | .8036 |
| 24 | .9719 | 49 | .9362 | 74 | .8799 | 99 | .7989 |
| 25 | .9708 | 50 | .9343 | 75 | .8773 | 100 | .7939 |

Figures given as percent by volume @ 68°F, per U.S. Bureau of Standards.

**CHART 2-3**

## $CH_3CH_2OH$ SPECIFIC GRAVITY CORRECTION FACTORS FOR TEMPERATURES OTHER THAN 68°F (20°C)

| °C | °F | S.G.* × | °C | °F | S.G.* × | °C | °F | S.G.* × |
|---|---|---|---|---|---|---|---|---|
| 0 | 32 | 0.9792 | 15 | 59 | 0.9948 | 30 | 86 | 1.0103 |
| | 33 | 0.9797 | | 60 | 0.9953 | | 87 | 1.0108 |
| 1 | 34 | 0.9803 | 16 | 61 | 0.9959 | 31 | 88 | 1.0114 |
| | 35 | 0.9809 | | 62 | 0.9965 | | 89 | 1.0120 |
| 2 | 36 | 0.9814 | 17 | 63 | 0.9970 | 32 | 90 | 1.0126 |
| 3 | 37 | 0.9820 | 18 | 64 | 0.9976 | 33 | 91 | 1.0132 |
| | 38 | 0.9826 | | 65 | 0.9982 | | 92 | 1.0138 |
| 4 | 39 | 0.9831 | 19 | 66 | 0.9988 | 34 | 93 | 1.0143 |
| | 40 | 0.9837 | | 67 | 0.9994 | | 94 | 1.0149 |
| 5 | 41 | 0.9843 | 20 | 68 | 1.0000 | 35 | 95 | 1.0155 |
| | 42 | 0.9849 | | 69 | 1.0005 | | 96 | 1.0160 |
| 6 | 43 | 0.9855 | 21 | 70 | 1.0011 | 36 | 97 | 1.0166 |
| | 44 | 0.9861 | | 71 | 1.0017 | | 98 | 1.0172 |
| 7 | 45 | 0.9866 | 22 | 72 | 1.0022 | 37 | 99 | 1.0177 |
| 8 | 46 | 0.9872 | 23 | 73 | 1.0028 | 38 | 100 | 1.0183 |
| | 47 | 0.9879 | | 74 | 1.0034 | | 101 | 1.0189 |
| 9 | 48 | 0.9884 | 24 | 75 | 1.0039 | 39 | 102 | 1.0195 |
| | 49 | 0.9890 | | 76 | 1.0045 | | 103 | 1.0201 |
| 10 | 50 | 0.9896 | 25 | 77 | 1.0051 | 40 | 104 | 1.0207 |
| | 51 | 0.9901 | | 78 | 1.0057 | | 105 | 1.0212 |
| 11 | 52 | 0.9907 | 26 | 79 | 1.0063 | 41 | 106 | 1.0218 |
| | 53 | 0.9913 | | 80 | 1.0069 | | 107 | 1.0224 |
| 12 | 54 | 0.9919 | 27 | 81 | 1.0074 | 42 | 108 | 1.0229 |
| 13 | 55 | 0.9925 | 28 | 82 | 1.0080 | 43 | 109 | 1.0235 |
| | 56 | 0.9931 | | 83 | 1.0086 | | 110 | 1.0241 |
| 14 | 57 | 0.9936 | 29 | 84 | 1.0091 | 44 | 111 | 1.0246 |
| | 58 | 0.9942 | | 85 | 1.0097 | | 112 | 1.0252 |

*Specific Gravity

**CHART 2-4**

## AVERAGE YIELD[1] OF 99.5 PERCENT ALCOHOL PER TON[2]

| Material | Gallons |
|---|---|
| Wheat (all varieties) | 85.0 |
| Corn | 84.0 |
| Buckwheat | 83.4 |
| Raisins | 81.4 |
| Grain sorghum | 79.5 |
| Rice, rough | 79.5 |
| Barley | 79.2 |
| Dates, dry | 79.0 |
| Rye | 78.8 |
| Prunes, dry | 72.0 |
| Molasses, blackstrap | 70.4 |
| Sorghum cane | 70.4 |
| Oats | 63.6 |
| Figs, dry | 59.0 |
| Sweet potatoes | 34.2 |
| Yams | 27.3 |
| Potatoes | 22.9 |
| Sugar beets | 22.1 |
| Figs, fresh | 21.0 |
| Jerusalem artichokes | 20.0 |
| Pineapples | 15.6 |
| Sugar cane | 15.2 |
| Grapes (all varieties) | 15.1 |
| Apples | 14.4 |
| Apricots | 13.6 |
| Pears | 11.5 |
| Peaches | 11.5 |
| Plums (nonprunes) | 10.9 |
| Carrots | 9.8 |

[1] Probable yield from a short ton of the raw material, calculated from the average fermentable content.

[2] Jacobs, P.B., and H.P. Newton. *U.S. Dept. Agr., Misc. Pub. 327.* December 1938.

**CHART 2-5**

## AVERAGE YIELD OF 99.5 PERCENT ALCOHOL PER ACRE[1]

| Material | Gallons |
|---|---|
| Sugar cane (Hawaii, 18 to 22 months) | 889.0 |
| Sugar beets | 287.0 |
| Sugar cane (Louisiana) | 268.0 |
| Jerusalem artichokes | 180.0 |
| Potatoes | 178.0 |
| Sweet potatoes | 141.0 |
| Apples | 140.0 |
| Dates, dry | 126.0 |
| Carrots | 121.0 |
| Raisins | 101.7 |
| Yams | 94.0 |
| Grapes (all varieties) | 90.4 |
| Corn | 88.8 |
| Peaches | 84.0 |
| Prunes, dry | 82.8 |
| Pineapples | 78.0 |
| Rice, rough | 65.6 |
| Pears | 49.3 |
| Barley | 47.9 |
| Molasses, blackstrap | 45.0 |
| Apricots | 41.0 |
| Oats | 36.3 |
| Grain sorghum | 35.5 |
| Buckwheat | 34.2 |
| Wheat (all varieties) | 33.0 |
| Figs, fresh | 31.5 |
| Figs, dry | 29.5 |
| Sorghum cane | 26.4 |
| Rye | 23.8 |
| Plums (nonprunes) | 21.8 |

[1] Jacobs, P.B., and H.P. Newton. *U.S. Dept. Agr., Misc. Pub. 327.* December 1938.

**CHART 2-6**

be considered when selecting a source material for ethanol production. Availability, cost, and ease of conversion all enter into the choice. Some of the most common source materials are discussed below.

**Grains**—Corn, wheat, oats, rye, barley, and rice are the grains used most often in the production of ethanol in this country. Since they are primarily starches, grains must be milled, diluted, cooked, and treated with enzymes (or malted) prior to fermentation. Grains typically contain 50–65% convertible starch and sugar, and yield about 60–80 gallons per ton. Though ethanol extraction from grains *does* require the extra step of starch-to-sugar conversion, higher yields and lower costs usually make such a procedure worthwhile.

**Potatoes**—Tubers must be cut up or shredded, slightly diluted, and either steamed or boiled before enzymes are added to effect a starch-to-sugar conversion. A ton of regular potatoes will yield approximately 20–25 gallons of ethanol, while sweet potatoes yield as much as 35 gallons. (Damaged or sprouted potatoes *may* be used.)

**Fruits**—Because fruits are *saccharide* (sugar-containing) materials, they require no starch-to-sugar conversion step. Generally they are pressed or crushed to extract the sugary juice. Fruits vary widely in sugar content (see Chart 2-7) and therefore in yield . . . as well as in percentage of extraction (the amount of juice which can be extracted with reasonable effort).

Rates of extraction range from about 90% for "soft" fruits without much internal solidity (like melons), to approximately 75% for "hard" fruits (like apples and pears). Grapes produce the highest yield for a nondried, commonly available fruit . . . but then, when grapes are fermented, it's not usually—at least not yet—with an eye toward distilling the juice into liquid fuel for engines and heaters.

**Molasses**—A residue from the manufacture of sugar from cane and beets, molasses is perhaps one of the most commonly used raw materials in the commercial manufacture of ethanol. With a sugar content of 50–55%, it is an excellent source material . . . but it's generally unavailable (at least in quantity) to the small distiller.

**Sugar Beets**—With a sugar content of about 15%, sugar beets can be expected to produce about 20 gallons of ethanol per ton. The beets must be

| Fruit | Average Sugar Content |
|---|---|
| Grapes (all varieties) | 15% |
| Apples | 12% |
| Pears | 10% |
| Peaches | 8% |
| Oranges | 5% |
| Watermelons | 3% |

CHART 2–7

pressed or crushed, and—in the latter case—enzymes are usually added to the mixture to aid in liquefaction.

**Whey**—A residue of the cheesemaking process, whey is a good source of ethanol. However the process required to separate out the sugars is especially complicated if the whey proteins and other byproducts are not to be lost.

Actually, *any* starch- or sugar-containing material may be used to produce ethanol, but it is important to know the percentage of sugar content of such foods in order to decide if their fermentation and distillation are worthwhile. Extraction and processing requirements are other factors that should be considered in deciding if a material is practical and economical for use in ethanol production. (You can obtain such information from the reference works listed in the back of this text or from your local agricultural service.)

Regardless of the source material or materials that you use to make your alcohol, the ultimate success of your fuel-distilling venture will depend upon close attention to the following factors:

- Efficiency of the preliminary treatment (milling, pressing, etc.)
- Dilution to the proper sugar concentration
- Correct pH control (when applicable)
- Correct temperature control
- Use of correct enzymes (or malting)
- Cleanliness of equipment and bacterial control
- Use of an active, alcohol-tolerant yeast
- Prompt distillation

## ETHANOL FROM CELLULOSIC WASTE MATERIALS

Ethanol *can* be distilled from waste materials . . . including wood chips, bagasse (cane stalks), corn stalks and cobs, hulls, and even some municipal and industrial wastes. However, the process necessary to convert the cellulose contained in such materials to sugars is—at least at this point—both complex *and* technology intensive. Experiments are currently being conducted using a fungus mutant which is said to convert cellulose into glucose "almost instantly" . . . but until more work has been done in this area, cellulose-to-sugar conversion is not a practical process for the small-scale distiller.

# 3 THE BARE-BONES PROCESS

## THE DISTILLER'S ART

People have been homebrewing alcohol in this country for hundreds of years, primarily in the form of drinking liquor or "moonshine". While it's true that alky has been distilled for other purposes too, it's also true that when the *individual* crafts-(wo)man or entrepreneur set out to build a still, it was usually with the idea of cooking up a batch of homemade whiskey.

Of course, home "booze" production wasn't always illegal. Thomas Jefferson repealed the "infernal" tax on whiskey as one of the first acts of his presidency, and except for one short period during the War of 1812, the making of untaxed "likker" was legal in the United States until 1862! The Civil War was the signal for the reimposition of taxes on the manufacture and sale of alcohol . . . in the form of one of those "wartime emergency measures" that somehow never seem to get repealed.

Today, the making of "moonshine" *is* illegal, and the operation of *any* still must be regulated by the U.S. Treasury Department. (Such problems will be addressed more thoroughly in the legal issues section of this book.)

But, though making homebrewed alcohol for human consumption is not lawful for the small-scale home distiller, the technology which evolved over years of backwoods joy-juice manufacturing is a good place to start when describing the *basic* process which the individual home distiller has to go through in the production of alcohol fuel.

## DOWN ON THE FARM

When frontier Americans set up home stills to turn their corn into alcohol, it wasn't because such men and women were a bunch of whiskey-loving sots who preferred drinking to eating . . . it was because the basic facts of economics and transportation in those times made distillation of the farm product the most *sensible* thing to do! At a point in our history when taking one's corn crop to market meant days (sometimes weeks) of hard travel over rough backwoods roads and trails (during which the fresh-picked corn would be molding and mildewing and generally losing its value) . . . it simply made good sense to *distill* the corn into a product that *wouldn't* spoil, and which was easier to transport because it was more concentrated.

Today's particular problems of economics and

*Anyone interested in making fuel alcohol on a small scale needn't go beyond "55-gallon drum" technology. A clean container, a source of heat, and a canoe paddle are a good start.*

transportation have combined in a *different* way to make the home distilling of alcohol *fuels* just as sensible as whiskey-making was to those frontier farmers two hundred years ago. And ethanol production *can* be just as legal now as it was then! But heed a word to the wise: *Don't even think about using your brand-new still for brewing up a little 'shine between fuel runs.* The federal government takes a dim view of such goings-on, and the penalties are very stiff.

## A SHORT STORY ABOUT ALCOHOL

How important is it to know the technical data about the production of alcohol? As you'd probably imagine, most moonshiners knew little—if anything!—about the internal workings of starch, sugar, and ethanol. But then, their yield of alky—per bushel of corn—was small compared to the potential that the grain contained, and they often "boosted" the mix by adding large quantities of processed sugar . . . a practice that's become especially prevalent among 'shiners in modern times.

Some experienced and sophisticated twentieth century bootleggers *claim* that they are able to get up to three gallons of whiskey per bushel of corn with their crude stills, but such operations *do* boost with sugar, and their product only averages out to between 130 and 160 proof . . . a good deal less than the 180-proof juice we're aiming for to make first-rate fuel. In any case, the *average* backwoods entrepreneur gets an alcohol yield that's considerably less than three gallons a bushel. To produce the most from *your* efforts and money, you have to *know* what's going on (and why it's happening) inside your mash cooker, fermentation tub, and still!

Broadly speaking, the production of alcohol begins with sugar. If you dilute that sugar in solution with water and add yeast, it will produce ethanol . . . naturally. Unfortunately, the most abundant (and least expensive) sugar available for making alcohol is in the form of an insoluble matter called starch . . . which is actually a complex sugar technically known as *polysaccharide* (literally: "many sugars"). Fortunately, starch (actually a chemical linkage of hundreds of simple sugars) can be converted to a soluble sugar relatively easily. This conversion process is called *saccharification.*

Of course, some substances (such as fruit) already include the simple sugars which are susceptible to fermentation. However, these crops are generally more expensive to obtain, and do not contain the amount of alcohol-producing sugar per ton that starch-containing grains do. So, while using fruit could cut down on the steps required to make alcohol, it will rarely be cost effective (unless you have ready access to spoiled produce or canning factory wastes).

The substances responsible for the starch-to-sugar conversion are called *enzymes*. They are protein biocatalysts, which means that they're produced by living cells and can cause chemical change. There are thousands of enzymes, but they're so specialized that each one can often bring about only one kind of reaction. (For example, the enzyme *invertase* will act only upon the sugar sucrose—reducing it to glucose and fructose—and will *not* act upon any other compound sugars.) Because of such specialization, the use of proper enzymes for specific tasks is crucial to the successful distillation of alcohol. (The enzymes we will later recommend have been developed for and proven in the distilling industry. You'll read all about them in the next chapter.)

Enzymes *cannot act* upon unprocessed, hard (or "flinty") starch. First, the starch in the shell of the dried corn kernel *and* its internal cellular structure have to be broken down. This means that the corn must be milled and cooked until the substrates (the substances upon which the enzymes act) gelatinize . . . a process which is called *liquefaction*. Specific enzymes have been developed to aid in this task.

There are a couple of different ways to get the enzymes you need to prepare your mash. Old-time distillers "malted" their grain, a process in which grain seed (barley is preferred) is sprouted and mixed with the mash batch. This natural process produces what are called *cereal enzymes* . . . enzymes which are of the type needed to convert starch into a form of fermentable sugar, principally maltose. A more modern—and easier—method is to add certain commercially produced microbial enzymes of the *alpha-amylase* and *glucoamylase* varieties. These do a thorough job of converting starches to sugars, are heat tolerant, have a long shelf life, and have been—as we've pointed out—specifically concocted for alcohol fuel production.

After mashing, yeast is added to the *wort* (a brewer's term for unfermented mash). The yeast also produces some conversion enzymes—invertase, for

*Commercially produced powders—such as those pictured here—do a thorough job of converting starches to sugars, are heat tolerant, have a long shelf life, and have been specifically manufactured for home and industrial alcohol fuel producers.*

example—but, more important, it is the microbe that accomplishes the transformation of sugar to alcohol . . . the process known as *fermentation*.

The final products of the yeast conversion are ethyl alcohol and carbon dioxide. The chemical reaction goes like this:

$$\underset{\text{(sugar)}}{C_6H_{12}O_6} \longrightarrow \underset{\text{(carbon dioxide)}}{2CO_2} + \underset{\text{(alcohol)}}{2C_2H_6O\ (2C_2H_5OH)}$$

At this point the fermented mash (which is usually called "beer" because—with some recipe changes and without some of the fining steps—it is almost identical to the beverage, beer) is an alcohol-containing mixture. However, the concentration of alcohol in the beer is very low . . . anywhere from 8 to 16% depending upon a variety of factors. In *raising* the alcohol percentage, the final technique—called *distillation*—is used.

Distillation is simply the *concentration* of the alcohol already present in a solution. The process is based upon the fact that the boiling (i.e., the vaporization) temperature of alcohol is 173°F, whereas that of water is 212°F. To extract the pure alcohol from a solution of alcohol and water (which, along with the solids, is what the beer consists of), it is only necessary to bring the temperature of the beer up to a point above 173° but below 212°. (The preferred temperature is around 196°F.) The vapor is then collected and cooled to change it back into a liquid. If the process has been properly applied, the resulting liquid will be *high-proof* alcohol.

If that procedure sounds complicated . . . rest assured that it really isn't. People have been producing alcohol—of various grades and for many purposes—for thousands of years. It's only when you must get your proof absolutely *right* that the process can be a mite tricky. In the rest of this section we'll take you through a step-by-step introduction to the basics of alcohol production and the equipment needed for that production. Then, in the following chapters, we'll cover—in detail—the steps and equipment needed to turn corn, or whatever you're using, into high-grade fuel. Just remember: With the proper apparatus and attention to detail, the production of fuel-grade alcohol isn't difficult . . . it's just *exacting*. The more care you take in each individual step of the process, the greater chance you'll have of producing the fuel you're after. (For a flow diagram of ethyl alcohol production, see Fig. 3-1.)

## HOW IT'S DONE . . . AN OVERVIEW

Never forget that the production of alcohol is a *natural* process. The distiller's art consists mainly of arranging circumstances and helping nature, in order to get larger yields—of a higher-quality product—than would otherwise be possible. Lots of folks have made their own beer and wine . . . through fermentation, a simple process in which the natural sugars in fruits or grains are fermented to produce a low-grade—usually about 12% (24-proof) or less—alcoholic beverage. It is the *distillation* of the fermented solution—into the approximately 180-proof (90% pure) alcohol fuel that *you* should be aiming to produce—which requires a bit of skill.

Because corn is one of the most easily obtainable raw materials which can be used in the production of ethyl alcohol, and because it produces a relatively high gallonage per ton (see Chart 2-5), we'll use it as our natural "source" of alcohol for the purpose of describing production methods. *Unless it is specifically stated that some other raw material is being discussed, we'll be talking about corn from here on.*

## THE BASIC STEPS OF ALCOHOL PRODUCTION

[1] MILLING: The *starch source is ground* into a fine meal (but is still far *coarser* than commercial cornmeal). This procedure exposes the starch granules and permits suspension and dispersion in the following step.

[2] SLURRYING: *Water is added* to the meal to form a mash. The amount of liquid necessary varies, depending upon the sugar content of the source materials . . . since, when yeast is added later to begin the conversion, too great a concentration of sugar (which will soon result in a heavy alcohol percentage) will kill it. A final concentration of 10–18% starch or sugar solids will be the most efficient. The sugar or starch content of a material can be determined by laboratory analysis, by the use of reference works, or with a "balling" hydrometer. If you're not using a specific recipe, make a small trial run to determine the best slurry mix. Next, the

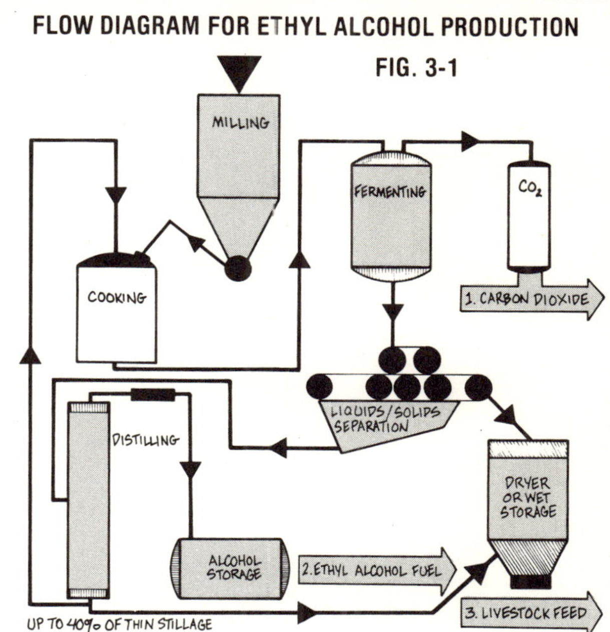


FLOW DIAGRAM FOR ETHYL ALCOHOL PRODUCTION

FIG. 3-1

### PRODUCTS FROM ETHYL ALCOHOL FERMENTATION

**1. $CO_2$**

One-half of the fermenting sugar is converted to carbon dioxide. It can be used for industrial application or in greenhouses for increasing plant growth.

**2. Alcohol**

The other half of the fermenting sugar is converted to ethyl alcohol. Since it contains fusel oil, esters, and aldehydes, it is not good for drinking but is a good fuel.

**3. Distiller's Grain**

Nearly all the protein is left in the solids, so distiller's grain becomes a high-quality feed for livestock. Protein is 28-30%, fiber is 12-13%, and moisture is 8-12%. Use it as a supplement to increase the protein in other feeds.

*FAR ABOVE: The milling process prepares the feedstock for eventual enzymatic breakdown. ABOVE: The ground meal is mixed with an appropriate amount of water to form a slurry, a cooking enzyme is added, and the blend is heated and agitated.*

mixture's *pH is adjusted*, if necessary (see the following chapter for details), and *enzymes* selected to aid in liquefaction are added.

[3] LIQUEFACTION: The *mash is heated* to gelatinize the starch and render it susceptible to enzyme breakdown. By this process the starch is converted into soluble, high-molecular-weight sugars called dextrins.

[4] CONVERSION: The *mash is cooled* to conversion temperature, the *pH is adjusted* once more, and the conversion *enzymes* are added. The mash is *held at this temperature* long enough to permit some of the dextrins to be converted into fermentable sugars. It is then *cooled* down to fermentation temperature.

[5] FERMENTATION: *Yeast is added* and the *temperature* of the mash *is held stable* as fermentation takes place. During this time the conversion enzymes continue to break down dextrins to fermentable sugars, while the yeast converts the sugars to ethanol and $CO_2$. Fermentation produces heat, so more *cooling may be necessary*—during this process—to maintain the proper temperature for yeast survival. When the fermentation is complete, the *solids are separated* from the solution by the process of filtration.

[6] DISTILLATION: The fermented *mash*—or "beer"—is *heated* to vaporize the alcohol, and the *vapors are collected and cooled* to condense the alcohol *back* to a liquid state. The remains from this process contain the residual grain, spent yeast, and water. The solids (DDGS) that are separated from the "beer" (both by distillation and—as mentioned in the preceding step—by filtration) can be used in animal feeds.

The above breakdown constitutes a very general description of the alcohol-making process. Specifics, such as temperatures, times of operations, types of enzymes and yeasts, pH levels, etc. will be dealt with in the following chapters.

## THE BASIC APPARATUS

Later on in this book we'll go into some detail about the different kinds of stills you can build, and we'll provide diagrams of MOTHER's woodburning stills. But before you build a distillation apparatus, you've got to have something to cook in it. And—while the equipment needed to produce a fermented beer isn't as technically complex as a still is—it's vital to the success of your operation.

The apparatus that you use for mixing, mashing, and fermenting your beer can be small and simple or large and somewhat complex. The size will depend upon such factors as how serious you are about making ethanol and how much you intend to make. In the chapter on stills, we'll be discussing both a little stovetop cooker and a farm-based monster still that produces several hundred gallons of alcohol per run. Most folks will want to make up more ethanol than the couple of quarts the small unit's capable of . . . but won't have the resources (or any need) for the large still.

Therefore, what we're going to do now is describe a *basic* setup which seems likely to meet the needs of most people who'll read this book. As we said, the equipment's uncomplicated . . . so if you want more or less capacity than our examples provide, it should be easy enough for you to design your own, once you've grasped the principles. We've broken the requirements down according to the same basic steps as those we used to illustrate alcohol production, for ease of understanding.

[1] MILLING: Unless you can buy your grain premilled or take it to a miller to be ground, you'll need a *coarse grinder* or *hammermill* for this job.

[2] SLURRYING: A *55-gallon drum* is great for this part of the process . . . *if* you can get one that's lined with fiberglass or plastic. Otherwise, you'll have to use plastic or wooden containers and then pour your slurry into a separate cooker for the next step. If you can get hold of a lined 55-gallon drum, it can double as your slurrying (mixing) vat *and* your cooker. To make it easier to transfer the cooked mash to your fermenting vat, it's also nice to have a valve and pipe runout set into the bottom of the barrel. Otherwise the mash must be dipped or poured from the cooker into the fermenter.

[3] LIQUEFACTION (cooking): A *cooker* is simply a *vat with a firebox underneath*, though of course you can work up a separate unit if it suits your needs. A firebox that'll burn wood, coal, or field waste (corn stalks and cobs, wheat straw, etc.) is ideal as the heating setup for cooking your mash. Remember: We're making liquid fuel because it's scarce, so there isn't really much sense to burning such fuel as a heat source to *make* ethanol.

[4] CONVERSION: This takes place during the cooking step, in the *cooker*.

*FAR ABOVE: The starch source can be coarse ground at a grain mill . . . or you can process the material at your own alcohol production site using a hammermill with a 3/16" screen. ABOVE: The meal should be added slowly to prevent lumping.*

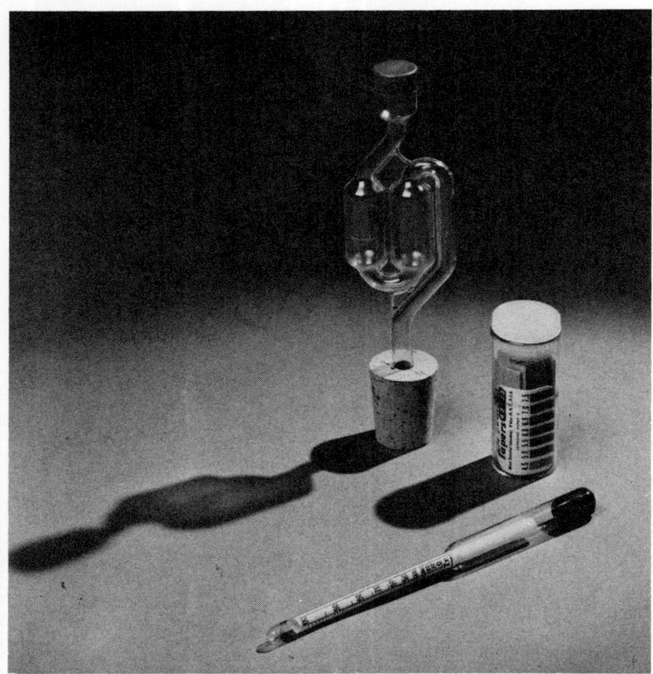

*You'll need a fermentation lock (to allow the $CO_2$ to escape from the fermenting vat), litmus paper (for testing the pH of the slurry), and a hydrometer (for determining the alcohol content of your finished product) to operate your home distillery.*

[5] FERMENTATION: Unless you've got a very steady source of lined 55-gallon drums, we suggest that the best small-scale way to ferment your cooked mash is in plastic garbage cans. They're light, easy to clean, and inexpensive . . . and allow for dispersal of your run (which means that if something goes wrong during the fermenting stage you won't be as likely to lose the entire batch).

[6] DISTILLATION: A still is simply a sealed cooking pot . . . like a pressure cooker, except that the vapors caused by the cooking of the contents are captured and routed elsewhere. (Stills will be discussed in detail in Chapter 6.)

Besides such large pieces of equipment and the basic raw materials, you may need the following:

**A Hydrometer**—for testing the alcohol content of your output (see the product source list at the end of this book).

**Fermentation Lock(s)**—to allow the escape of $CO_2$ from the fermentation vat *and* to give you an idea of when the ferment has stopped (see the product source list at the end of this book).

**Litmus Paper**—for testing pH (available through laboratory supply stores, etc).

**Sulfuric Acid and Agricultural Lime** (or their equivalents)—for rectifying overacid or overalkaline states in the batch (available through garden supply stores, etc.).

## STEP-BY-STEP (CAREFULLY)

Concocting a successful mash is a lot like accomplishing *any* worthwhile project . . . you have to be careful. If you use a good recipe and the right tools, *and* take a few basic precautions, the "brew" will turn out just fine. But if you try to cut corners—in any of these three areas—you'll just be wasting your time and money!

In this chapter we're going to present the recipe that MOTHER's researchers have found will give you the *most* fuel yield for your dollar . . . with the *least* fuss. And—though we're using a corn-based process—we'll throw in some tips about how to make ethanol from other bases, as well. But first, we're going to do a little talking about how and why a mash happens . . . that is, what it is that occurs in the vat to turn starch into sugar and sugar into alcohol. If you have a basic understanding of what's going on in your mash tub, you'll be less likely to make a time- and money-wasting mistake.

## ENZYMES

As we mentioned when describing the basic alcohol-making process in the previous chapter, the key step—in the conversion of the starch in corn into sugar—is the addition of enzymes. The enzymes act upon starch to convert it into simple fermentable sugars such as maltose and glucose, a necessary process in the making of alcohol. Some home distillers will be using materials which don't need to go through this step because the substances *already* contain fermentable sugars . . . some examples are fruit squeezings, molasses, and whey. But most ethanol manufacturers are likely to begin with starchy substances such as corn, wheat, tubers, and so forth . . . and will have to *convert* the starches that these products contain.

There are several different ways to go about the conversion, but all of the various methods require the same substance . . . an enzyme called *amylase*. Amylase preparations are the biological catalysts which can *break* the bonds of the starch molecule . . . freeing the sugar molecules of which it is composed. In practical terms, this means the conversion of a substance from starch to sugar (a transformation that we've already identified as *saccharification*). The enzymes that perform this necessary function are contained both in malted (or sprouted) grains and in commercially produced

*Alpha- and beta-amylase were produced by malting (sprouting) before the enzymes became commercially available.*

THE CONVERSION OF STARCH TO SUGAR TO ALCOHOL
FIG. 4-1

STARCH
HEATING CAUSES GELATINIZATION.
GRANULES RUPTURE.
ENZYME ALPHA-AMYLASE [HYDROLYSIS]
DEXTRINS
ENZYME BETA-AMYLASE OR GLUCOAMYLASE [HYDROLYSIS]
MALTOSE, GLUCOSE, OR FRUCTOSE
YEAST SACCHAROMYCES CEREVISIAE
$CO_2$
ALCOHOL
ZYMASE

and purified enzyme solutions. Whichever of these you choose to use, it is enzymes that are doing the work of converting starch into sugar. And it's important to remember that these microbes are *alive*, and thus require a specific, stable niche if they're to remain viable. If improper procedures are used during the conversion process—if, for example, the mash becomes contaminated, or heat levels are too high—your alcohol yield can be adversely affected. *Correct and stringent sanitary procedures (and careful attention to heat levels) are necessary to prevent this loss of yield.*

MOTHER's Three-Step Mash Recipe, which will be detailed later in this chapter, calls for the use of two different amylase enzymes. The first is a food-grade bacterial *alpha-amylase*, which is capable of rapidly reducing the viscosity of gelatinous starches . . . converting them into dextrins. Because it can be used in starch liquefaction processes (such as brewing) which employ high temperatures, the alpha-amylase is a valuable tool whenever heat is used in the gelatinization of a raw material such as corn.

Our recipe also calls for a complicated *glucoamylase*—an inexpensive malt replacement developed by the neutral grain spirits industry—which actually breaks down the starch molecules, converts them into fermentable sugars, and denatures the mash. (See Fig. 4-1 for an illustration of the starch-to-sugar-to-alcohol breakdown.)

## MALTING

Malting (sprouting) your own grain is an inexpensive method of producing enzymes without having to buy a commercial preparation (either cereal *or* bacterial), but it *does* entail a lot of work. All starchy (cereal) grains produce both alpha- and beta-amylase enzymes when moistened and allowed to sprout, and the method of sprouting to produce enzymes was used by home distillers and brewers for *hundreds of years* before commercially produced cereal and bacterial enzymes became available. While barley has been found to produce the greatest quantity of enzymes by volume—making it the most economical grain to use—corn, wheat, rye, and other grains can also be sprouted. The thing to remember is that undried, or "green", sprouted grain will not keep and has to be used immediately. Because of the time and work involved in sprouting grain *and* the additional labor that's necessary to dry a portion to use in malting subsequent batches (so that you won't have to go through the sprouting process with every single batch of alcohol you make), most folks find it less costly in the long run to buy commercially malted grain or to use the bacterial enzymes mentioned above.

## YEAST

Just as enzymes are responsible for the conversion of starch to sugar, *yeast* converts sugar to alcohol. Yeast is a fungus that thrives in sugar solutions. When placed in a sweet liquid, the yeast "plant" goes through eleven separate conversion stages, secreting a complex of enzymes which are responsible for the production of alcohol from sugar. The whole process is called *zymolysis* . . . a term that refers to a procedure in which several enzymes and coenzymes actually perform the operation of removing an atom here and replacing it there. The result is the conversion of sugar to carbon dioxide and alcohol.

The point of this entire process—as far as the yeast is concerned—is reproduction. As they multiply, the tiny plants just happen to create the byproducts for which they are prized.

Yeast is added to bread dough so that the slight fermentation action which takes place in the presence of the sugar in the dough will produce carbon dioxide bubbles. As a result, the bread "rises", becoming lighter, airier, and more palatable. In fermenting solutions to produce alcohol, the carbon dioxide is usually "bubbled off" into the atmosphere . . . leaving only the ethanol and other liquids which make up the "beer". However, in large commercial distilling operations, provision is sometimes made to capture the carbon dioxide in separate holding tanks, thus allowing its resale. Doing this makes the production of alcohol from vegetable matter an even more economically feasible proposition.

Because yeast is a living organism, the environment in which it lives will have a great effect on how efficiently it does its job. In the absence of air (oxygen in the water of the slurry), a far greater portion of the sugar goes off as alcohol and carbon dioxide. On the other hand, if the mash is aerated (exposed to the air, by stirring, etc.), the yeast will produce *more* new yeast cells but *fewer* valuable "wastes". The lesson here is to leave the yeast

*A home-scale self-watering sprouting cabinet (the plans are available from Mother's Plans, P.O. Box A, East Flat Rock, North Carolina 28726) could be useful to a small-scale distiller.*

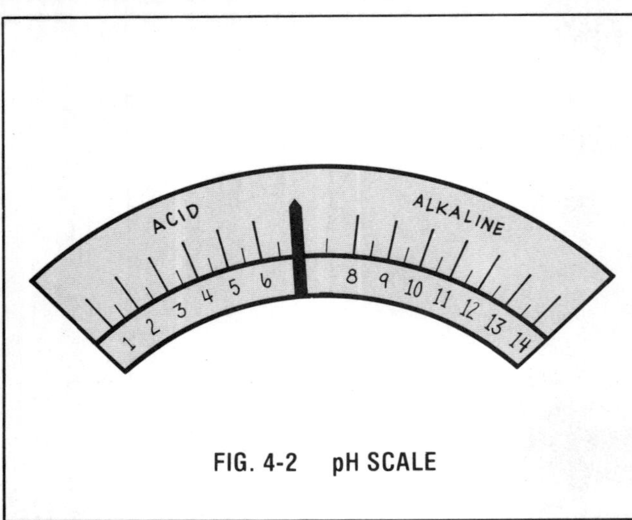


FIG. 4-2 pH SCALE

undisturbed and it will give a greater alcohol yield.

Chemicals or heavy metals can retard or stop yeast growth. It can also be adversely affected by heat or cold . . . the optimum temperature (which should be maintained during fermentation) is between 77 and 90°F.

Yeast plants can't go on making alcohol and carbon dioxide forever (if they could, there'd be no need for distilling to concentrate the ethanol). Yeast begins to become dormant when alcohol concentrations in the solution reach about 12% by volume, thus ending the sugar conversion process. It is at this point that the end product "beer" has been produced and that the next step, distilling, must begin.

## pH: WHAT IT IS AND HOW TO CONTROL IT

The pH of a solution is simply a measure of acidity or alkalinity, expressed on a scale of 1 to 14. Therefore, an absolutely neutral mixture will have a pH of 7, while pH readings of 1 through 6 indicate acid solutions, and readings of 8 through 14 indicate alkalinity (see Fig. 4-2). Various substances start with one pH reading but change as they are cooked or as additions are made to them. *Since the degree of alkalinity or acidity is crucial to such micro-organisms as enzymes and yeast, it's quite important that you monitor the pH level in your mash closely.*

At various points throughout the processes of mashing and fermenting, readings for pH must be taken and pH levels adjusted, should they be too alkaline or too acid. Readings can easily be taken with pH test papers (litmus papers) which are available from laboratory supply outlets, swimming pool supply houses, and garden supply houses. The pH level can be adjusted by adding sulfuric (or citric or hydrochloric) acid to lower the reading, or agricultural lime to raise it. (NOTE: The commercial enzymes used in MOTHER's Three-Step Mash Recipe don't require such adjustments.)

Most grain mashes tend to have a naturally acid pH, while saccharide mashes (those made from fruit, etc.) will usually be alkaline. Either extreme must be adjusted *after* testing.

## FERMENTATION

Fermentation (the conversion of sugar materials to alcohol) is simply the action of microscopic yeast fungi upon a sugar solution. The minute plants consume the sweet substance (which is either added to the mixture in the form of cane sugar, or "freed" from the corn itself by sprouting, or both) and give off carbon dioxide gas and alcohol.

When the yeast is added to the mash (the brew should be at a temperature between 70 and 90°F), fermentation begins. Any of a variety of yeasts can be used, all the way from Distiller's Active Dry Yeast (which is usually available only in large quantities) to baker's yeast (which can be obtained from bakery supply houses or, perhaps, from your local baker) to the powdered dry or cool cake yeasts you can find in any supermarket. (NOTE: MOTHER's Three-Step Mash Recipe includes the necessary yeast in the final enzyme addition.)

All you need to do after your brew has cooled properly and you've added the yeast is sit back and wait for it to ferment. But remember . . . one of the most important details in preparing *any* mash is to keep your containers and implements clean. The fermentation process depends upon the action of the yeast alone . . . and the presence of *outside* micro-organisms—such as those found in the air all around us—can render a whole batch useless.

Therefore, it's very important to use bleach-scrubbed (*and* thoroughly rinsed) plastic or wooden vessels for every step in the mixing and fermenting process. (Metal drums can also be used . . . *if* they're lined with fiberglass—or a plastic sack—and made to seal.)

Be sure to cover the barrel to keep air from the mash. (Not everyone does this, but it *does* improve the efficiency of fermentation . . . after all, we're trying to produce an ideal batch of alcohol, *not* an ideal environment for yeast to reproduce in record numbers.) Many distillers equip the lid or cover with some type of air or fermentation lock that lets the fermentation-produced $CO_2$ *out*, while not allowing any air *in*.

As we've pointed out, the mash should have cooled to a temperature of between 70 and 90°F before yeasting. After yeasting, fermentation begins and the little plants start to reproduce. At the *height* of fermentation the carbon dioxide gas which the yeast plants give off will actually create heat within the mash. The mixture will appear to boil at this stage, though of course that apparent "boiling" is brought about by escaping gas. (Still, additional heat *is* generated by the reaction.)

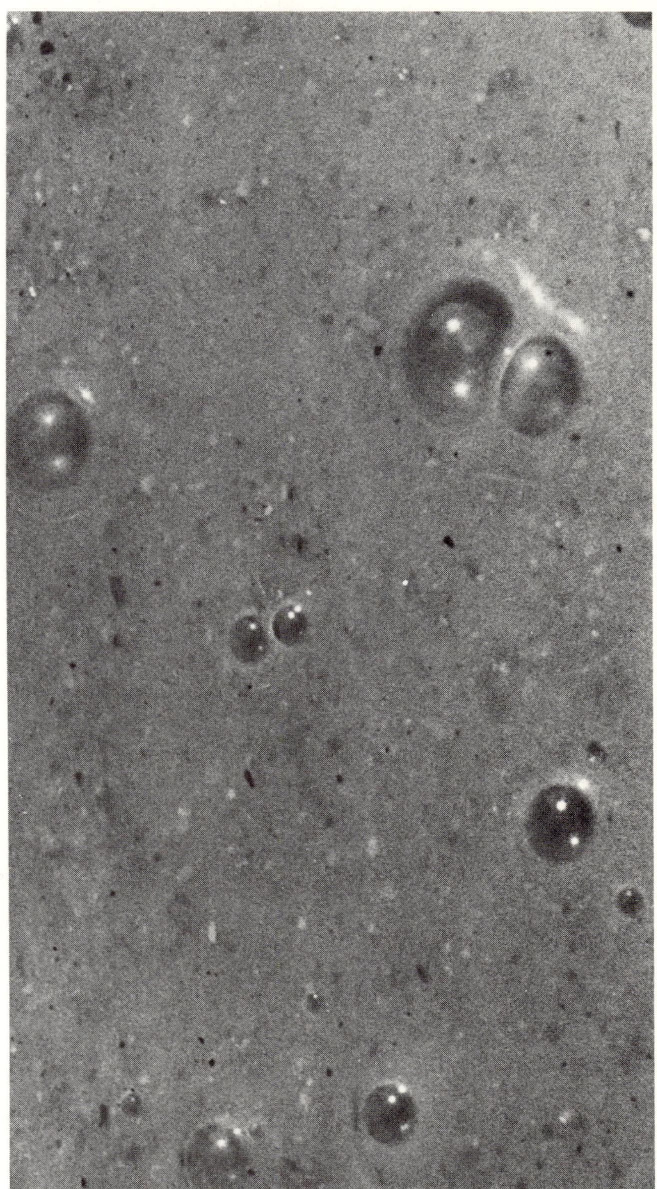

*As the yeast-inoculated mash mixture begins to ferment, the multiplying micro-organisms emit bubbles of $CO_2$. (It's important to use carefully cleaned containers during this stage.)*

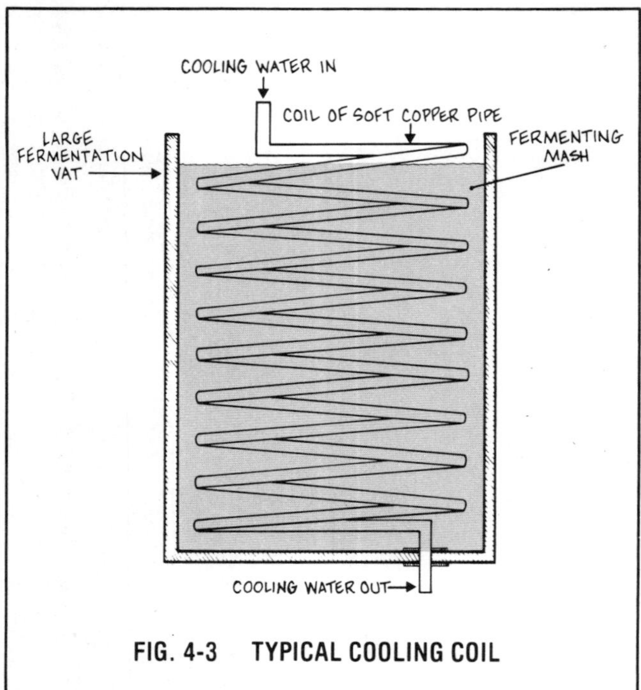


FIG. 4-3 TYPICAL COOLING COIL

Since you won't want to allow the temperature of the mash to rise above 90–95°F (remember that yeast plants can be killed or rendered dormant by extremes of temperature), it's important that you monitor your batch and cool the mixture . . . if you're preparing large enough batches of mash to produce a significant temperature change. While this can be done in a number of ways, the most efficient is to sink a cooling coil inside of the mash tub. MOTHER strongly recommends that you build yourself one of these devices if you're interested in large-scale alcohol production . . . using vats of around a 1,000-gallon capacity. (In addition, a cooling coil is necessary for MOTHER's Three-Step Mash Recipe.)

When the temperature of your fermenting mash starts to get too high, you can simply run cool water through the hoses, into the coil, and out again. It's as easy as that . . . except—when you're cooling your mash—be sure to monitor the temperature periodically. If you allow it to fall *below* 60°F, the yeast will become dormant and the fermentation will stop. (For an illustration of a cooling coil, see Fig. 4-3.)

If all this represents too much technology for you, you *can* lower the temperature of your batch by simply removing the cap from the fermentation tub and letting air cool the mash, but this will add a day or so to the fermenting process and (though it's unlikely) *may* allow harmful bacterial matter to get into the batch. Or, if you want a down-and-dirty method, you can suspend plastic trash bags full of ice in the mash. The trouble with this technique is that it's impossible to regulate, and inevitably will disturb the mash.

Fermentation should be complete in 2-1/2 to 3 days, if you're using MOTHER's Three-Step Mash Recipe (given in Chart 4-1). Watch the action of the bubbles in the air trap closely after the third day (but *don't* open the containers). When carbon dioxide gas no longer "perks" through the vent hole, your brew is ready. (Be careful, though: Mash *has* been known to stop "working" for a day . . . and then set to bubbling again with a vengeance.)

Another method of telling when your batch is ready is to wait for the "cap"—a layer of matter that will build on top of the mash—to drop to the bottom of the container. (Of course, this technique is best used with a transparent or translucent vessel.) When the mash has finished working, simply

*ABOVE, LEFT: Here a homemade "mash press" is being used to separate solids from a batch of mash prior to distillation. BELOW, LEFT AND ABOVE, RIGHT: The fermented grain residue that's separated out from the mash is very high in protein.*

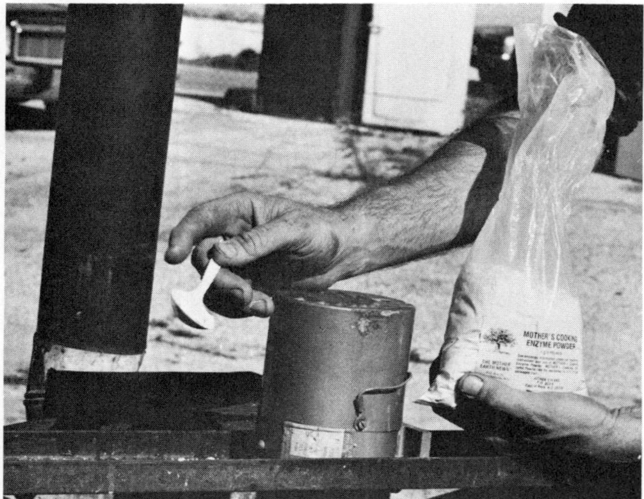


*TOP: A milled bushel of corn is slowly added to 30 gallons of water. The mixture is stirred while the grain is poured in. ABOVE: Once the corn is thoroughly blended, one ounce of MOTHER's Mash Cooking Enzyme (mixed in water) is added.*

uncap the container, strain the mixture through a few pieces of washed burlap, and pour the resulting pure mash into whatever still you're planning to use. You're now ready to begin your career as a distiller of fuel-grade alcohol.

## MOTHER'S THREE-STEP MASH RECIPE

Our successful ethanol production begins with a starch-rich concoction which converts, easily and thoroughly, to a yeast culture's favorite food: fer-

**CHART 4–1**

### MOTHER'S NEW, IMPROVED THREE-STEP MASHING RECIPE

#### MILLING

Shell, clean, and grind a bushel of corn (56 pounds) into a fine meal of about the size needed for livestock feed. Use a 3/16″ screen on a hammermill (or a similar grinder) to eliminate any large starch grains. However, *do not* grind the corn into a flour. If the grains are too small, it'll be very difficult to separate the solids from the mash . . . with a resulting loss of feed grain *and* a miserable mess inside your still.

#### STEP ONE: COOKING

Start with 30 gallons of water in your cooker, and then add the cornmeal slowly . . . to prevent lumping. Once the meal is mixed in, stir in approximately one ounce of MOTHER's Alcohol Fuel Mash Cooking Enzyme (an alpha-amylase) mixed in water, and bring the mixture up to 170°F (77°C). Hold the mash at this temperature for 15 minutes . . . stirring vigorously throughout the process. Then bring the liquid to a rapid rolling boil and hold it there for 30 minutes more. Be particularly careful that the mash doesn't stick to the bottom of the cooker. (For batches larger than a bushel, we recommend using an automatic agitator, which should spin at 30 to 45 RPM.)

#### STEP TWO: CONVERTING

Using the cooling coil, bring the temperature of the mash down to 170°F (77°C), and add another ounce of MOTHER's Cooking Enzyme (mixed in water). Keep the mixture at this temperature for 30 minutes, while you agitate it constantly.

mentable sugars. When the fungi find a generous supply of their preferred fare, they reward the distiller with a high rate of alcohol yield. MOTHER's researchers have been busy mashing, fermenting, and distilling various corn-based recipes ever since we first became interested in alcohol fuels. This experience, combined with some expert advice on fuel-grade alcohol preparation, has resulted in the recipe shown in Chart 4-1 . . . a formula which returns a maximum volume of flame-grade juice.

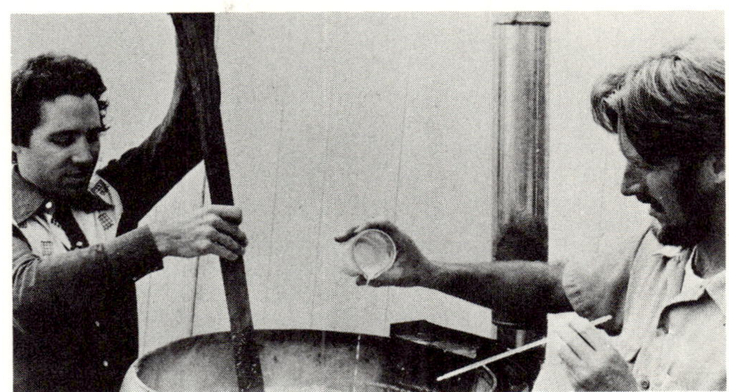

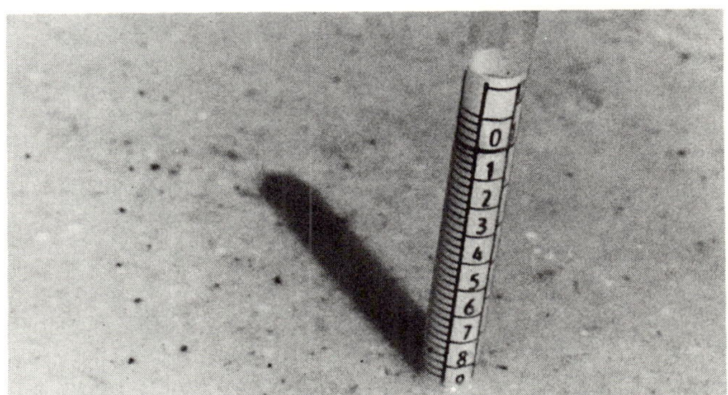


*TOP: After the mash has boiled, another ounce of cooking enzyme is added. CENTER: When the mash reaches 90°F, the fermentation powder is mixed in. ABOVE: A saccharometer (or triple-scale hydrometer) is used to measure specific gravity.*

### STEP THREE: FERMENTATION

Start cold water flowing through the cooling coil again, to reduce the temperature to 90°F (32°C) as rapidly as possible. Once the mash has cooled, add two ounces of MOTHER's Alcohol Fuel Fermentation Powder (a complex glucoamylase, yeast, *and* denaturant combination), stir the mash for 10 minutes, and then cover the tank.

While it's fermenting, the mash must be kept between 85 and 90°F (29–32°C). Consequently, you may need to cover the tank with wet burlap in hot weather . . . or insulate it during colder months. At this temperature, the mash will reach maturity in 2-1/2 to 3 days.

### TESTING PROCEDURES

USING A SACCHAROMETER: At the beginning of fermentation, the specific gravity of the mash should be about 1.080 (8 to 12% alcohol potential), while by the end of the process it will have dropped to 1.007 or less (0 to 1% alcohol potential). Once the specific gravity has remained constant for 6 hours, you can be sure that the mash is ready for distillation. But to double-check that complete conversion has been attained, both a standard starch test (using iodine) and a glucose test (using glucose test strips available at drug stores) must read negative.

### FOR THE THREE-STEP MASHING RECIPE, YOU MUST HAVE A COOLING COIL

To make a cooling coil, just wind a 30-foot length of soft copper tubing around a large pipe (6 inches, or more, in diameter), and add garden hose adapters at each end. Attach the hoses to the tube, and drop the assembly into your cooking vat.

## OTHER MATERIALS, OTHER METHODS

The recipe we've given you is the most efficient we've been able to come up with for the production of ethanol from corn. But—as we mentioned earlier—there are many other materials from which ethanol *can* be made.*

The saccharide products—such as fruit—are easy to work with (though not always available) because their carbohydrates are already in the form of sugar. Since there's no need for starch-to-sugar conversion, the ethanol-producing process is shortened and simplified considerably.

Before fruit is fermented, its sugar must be extracted . . . a process generally performed with a press such as is used to crush grapes, apples, and sugar cane for their juice. This is only partially effective, however, and leaves a significant sugar residue within the skins and pulp. This sugar can be extracted by adding a small amount of water to the residue, cooking the mash at a low temperature, and then re-pressing it.

Once the sugar juices have been extracted, the sugar concentration in a potential mash must be adjusted to suit the growth of yeast . . . between 14 and 18%. Such a concentration should be measured with a "balling" hydrometer—sometimes called a saccharometer—when the liquid is at a temperature of 60°F. Excessively high sugar concentrations (which inhibit yeast growth by promoting more rapid alcohol fermentation) should be diluted with water, whereas liquids with low readings (which are wasteful of fermenting space *and* the energy used in distillation) should be augmented with a concentrated sugar.

The extracted sugar juice (which *should* contain roughly 20% solid material) need only be adjusted for the proper pH balance (4.0–4.5) and "beefed up" with yeast nutrients (available through winemaking supply stores) . . . since sugar products don't contain the vitamins and minerals necessary for the yeast to do a good job.

Some of the more sophisticated distillers actually go so far as to augment their sugar-heavy mashes with nitrogen and phosphorus . . . two nutrients which sugars tend to be deficient in. Ammonium salts—such as ammonium sulfate or phosphate—are added to encourage a healthy yeast culture. Such a procedure is usually more complicated than the backyard alcohol producer should con-

*See Appendix

sider . . . but—in the event of yeast growth problems—some cautious experimentation might be in order.

Prepare your yeast by first mixing a yeast starter. Remove approximately 10 gallons of mash from the primary fermentation mixture—or "wort", as it is called in brewing technology—and stir 1/2 pound of dry active yeast into it. Then allow the starter to incubate for six hours . . . holding the temperature between 77 and 88°F. This technique promotes the rapid growth of yeast and speeds up the entire fermentation process.

Combining the yeast-inoculated mixture with the main mash is called "pitching". There are several ways in which the yeast starter can be added to the fermentation tank . . . but the most important aspect of the process is keeping the mash well aerated. One technique consists of using baffles which the mash flows over as it enters the tank. The splashing introduces air to the wort, which (as you now know) encourages a thriving yeast population. Another approach is to place a compressed-air line —with a bacterial filter—in the bottom of the tank. However, for the small producer, a human-powered stirrer (a canoe paddle, for example) will be satisfactory. (Of course, once the yeast population is thoroughly established, aeration must be halted to allow the microbes to adapt to the anaerobic conditions which result in maximum production of alcohol.)

Since the temperature of the fermenting mash should—ideally—be around 70°F, the pitching temperature will be determined by the ambient air temperature. For example, if the outside mercury is quite high, the initial temperature of the mash should be in the low sixties . . . and it may be necessary to cool the wort as the yeast begins to produce warmth of its own. On the other hand, low air temperature dictates an introductory level of 75 to 80°F. (Note: Any temperature *over* 95°F will only evaporate alcohol and encourage bacterial growth.)

Assuming that you have properly controlled the sugar concentration, the pH, the yeast nutrition, and the temperature . . . fermentation should be completed in about 50 hours. Activity will lessen in the mash, and the cap on top of the mixture will break apart and sink once the yeast has done its job.

If you have access to a large amount of molasses (which contains 60% sugar by weight), you'll find it

*Sugar cane (*Saccharum officinarum*) is a high-quality ethanol-producing crop. This seasonal Brazilian variety of the tropical grass can yield up to 268 gallons of 99.5% alcohol per acre.*

to be an excellent source of ethanol . . . in fact, because its sugar content is so high, molasses must be diluted with water (three parts water to form a 15% sugar solution). The pH is then adjusted, and yeast is added. Because molasses contains few of the nutrients needed by yeast in order to grow, "*backslopping*"—the addition of the dead yeast residues from previous batches—is advisable, in proportions of up to 40% of volume. This process can provide the yeast nutrients necessary in *fruit*-based mash, too. *[NOTE: In the whiskey industry, backslopping is the means by which that taste-distinctive "sour mash" bourbon is made . . . the slightly sour taste of the still residue gives the liquor a particular flavor.]*

Other saccharide products such as sugar beets, cane sorghum, and sugar cane can also be processed to make ethanol. Most require pressing or crushing, shredding, and then cooking to release the alcohol-producing sugar inside. However, as noted earlier, such materials produce considerably less alcohol per ton than do the starchy grains. (If you have access to large, inexpensive supplies of these products, though, they *can* be used very successfully as ethanol producers.)

In addition, there *are* a few other source materials in the *starch* family that you might want to consider. Potatoes contain between 15 and 18% fermentable material (sweet potatoes can have as much as 27–28%) and have been used as a source of alcohol for hundreds of years . . . especially in eastern Europe. On the average, a ton of spuds yields between 20 and 25 gallons of ethanol (sweet potatoes can produce as much as 35 gallons per ton). The tubers are usually cut up or shredded and then cooked with steam under pressure. If that procedure isn't practical—and it won't be for most home distillers—the potatoes can be covered with a minimum amount of water (they contain up to 80% water naturally) and cooked for several hours. (Sweet potatoes have less water . . . about 65% . . . and so will require more dilution.) The finished cooked product should be a uniform mash that looks much like a smooth potato soup . . . help it along, if you have to, by crushing the spuds with a large mashing mallet.

Potatoes are generally premalted (malt or enzymes are added to aid in the breaking down of the cellular structure during the liquefaction process). Then the pH is checked, conversion enzymes (or malt) added, and the mash stirred throughout the conversion process. Cooking and conversion times will vary according to circumstances, of course, and the use of a starch test to determine when the sugar concentration has reached a sufficiently high level is recommended. Following conversion, the pH is adjusted again, yeast is added, and the mash is allowed to ferment. If the potatoes have not been skinned, it will be necessary to strain the ferment prior to distillation.

## OTHER RECIPES

As we said, sooner or later every home distiller is probably going to start monkeying with his or her recipe, trying to come up with a more efficient way to produce alcohol. And that's good . . . experimentation is one way things get better. But *don't* make the mistake of using one of the mash recipes you might find in books about how they made whiskey and other liquors in the old (and not so old) days . . . and if a relative of yours used to be involved in such doings, don't ask him for his formulas, either. Moonshine recipes just aren't particularly efficient or cost effective, since the backwoods liquor runners had concerns that the alcohol fuel distiller doesn't have to worry about (such as taste and potability), and often used large amounts of sugar in their mashing recipes because they weren't converting their starches efficiently. Using sugar is not—in most cases—an economical way to produce alcohol fuel. And low-cost fuel is, of course, exactly what you're after.

*Jerusalem artichokes (*Helianthus tuberosus*) have a great potential as a source of alcohol. The tuber-bearing sunflowers are hardy, produce good yields, and require a minimum of care.*

## RECORD KEEPING

It's important to you—as a distiller—to keep records, both so that you'll be able to figure out what you did and why it did (or didn't) work, *and* so you can stay within the government's requirements. The basic kinds of record keeping follow, and working examples can be found in the Appendix.

**[1]** The **Alcohol Production Record Sheet** should help you keep track of your recipe *and* how efficient it is (almost every home distiller is going to end up doing a little experimenting on how to make his or her particular formula better). By keeping close track of *what* you used and *how* you used it, you'll be able to figure out what process gives you the most fuel-grade alcohol for your time and money.

[2] The **Alcohol Denature Record Sheet** is to help you satisfy the government's demands. Most fuel alcohol must be denatured . . . that is, made undrinkable through the addition of contaminants. (Of course, the use of MOTHER's mash fermentation enzyme powder provides a denatured mash.) However, the requirements for denaturing are now being *changed*. Consult your regional Bureau of Alcohol, Tobacco, and Firearms for details (see Chapter 8).

[3] The **Alcohol Disposal Record Sheet** will help you keep track of the uses to which you've put your alcohol fuel and how efficient it is for each purpose.

In addition, you may want to calculate the cost of the ethanol you produce. A sample form for figuring up the total expense is shown below.

**How to Figure the Cost of ETHANOL for a do-it-yourself distiller**

| | | |
|---|---|---|
| Raw Material[1] ................. | $______ | |
| By-Product Credit[2] ............. | ______ | (subtract) |
| Net Raw Material .............. | ______ | |
| Conversion Cost[3] .............. | ______ | |
| ETHANOL Cost ................ | ______ | (subtotal) |
| Equipment Cost[4] ............... | ______ | |
| Miscellaneous Cost[5] ........... | ______ | |
| | $______ | TOTAL |

[1] If you grind your own grain, DO NOT add the cost, but if it's done by a miller, DO add the cost.

[2] Distiller's grain has value. Subtract the market price if you're feeding it to your own livestock, or the price you receive if you're selling it to someone else.

[3] Cost of enzymes, yeast, and fuel for cooker/still.

[4] Investment in machinery (tractor/grinder, buildings, distillery equipment, etc.), but only that percentage used for actual ethanol production.

[5] Fuel for tractor, electricity, etc. used in ethanol production.

# 5 THE DISTILLATION PROCESS

## WHAT IT IS . . . HOW IT WORKS

A fully fermented mash mixture—when ready for distillation—is only between 20 and 25 proof . . . which is the equivalent of 10–12.5% ethanol. The process of distillation will drive the alcohol vapors *out of* the solution and allow them to condense in a purer (less diluted with water) form. In the case of ethanol distillation, then, we are extracting pure ethyl alcohol from fermented beer. By vaporizing the alcohol and driving it off as a gas, we can separate it from the water and solids which are the other constituents of the ferment. (For a diagram of the basic distillation process, see Fig. 5-1.)

As was mentioned earlier, distillation is possible because of the variance in the boiling (vaporization) points of alcohol and water. Pure 100% (200-proof) alky boils and vaporizes at exactly 172.9°F at sea level. Pure water boils at 212°F. (Chart 5-1 shows a boiling point diagram for ethanol/water solutions.)

The actual temperature at which a solution of alcohol and water will boil depends upon the ratio of the solution . . . that is, on what percentage of the mix is water, and what percentage is ethanol. The more alcohol present, the *lower* the temperature at which the solution will boil and begin to vaporize. Low-heat vapors will contain a greater concentration of alcohol than those produced at high temperatures, while the solution which is left behind will be more "watery". When the vapors are condensed to form a liquid, that solution will contain a higher percentage of alcohol than did the original ferment . . . but it will still contain *some* water. How high the water percentage is will vary inversely with how efficient the distillation process is . . . and that's largely a reflection of how effective the distillation apparatus, or still, is.

Let's take an example: In a simple stovetop test still, a solution which contains 10% alcohol and 90% water will boil at about 200°F and produce vapors which may contain only about 44% ethanol. Since that isn't nearly pure enough to be used as a liquid fuel, the recondensed vapors of such a first distillation must be *re*distilled. If the boiling point of the second distillation (or "run") is around 180°F, then the final product will be about 80% ethanol or 160 proof. To get the high-grade (around 90% alcohol) fuel which is our goal, we'd have to make several more runs, because each run

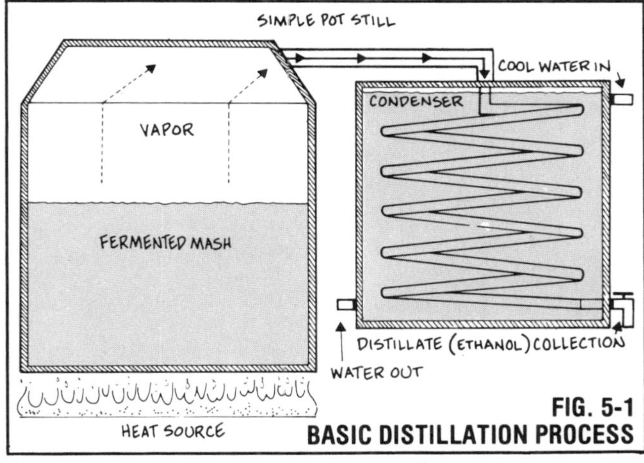


FIG. 5-1
BASIC DISTILLATION PROCESS

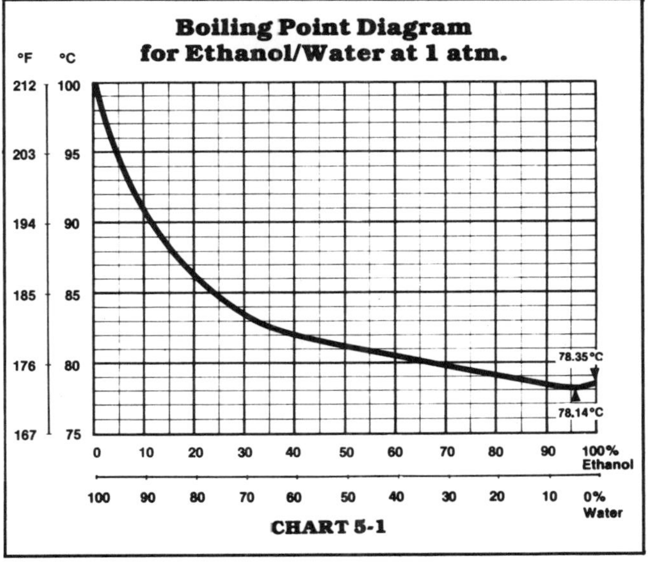


CHART 5-1

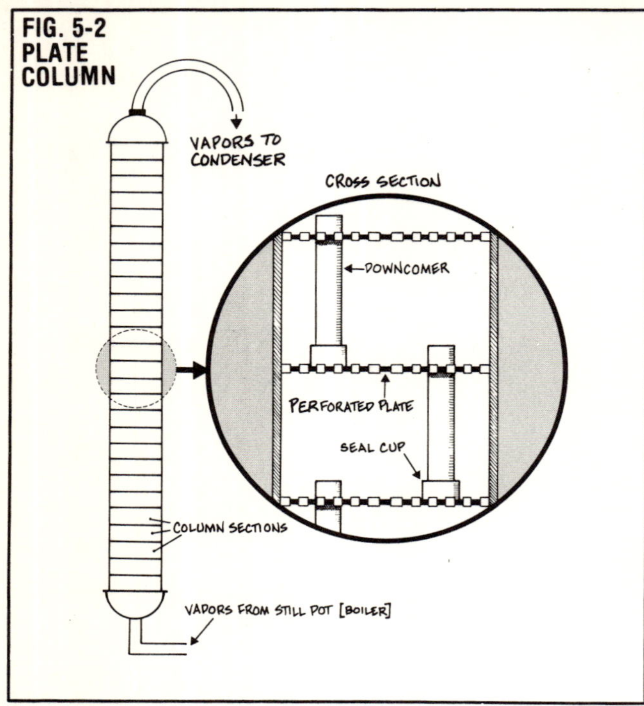


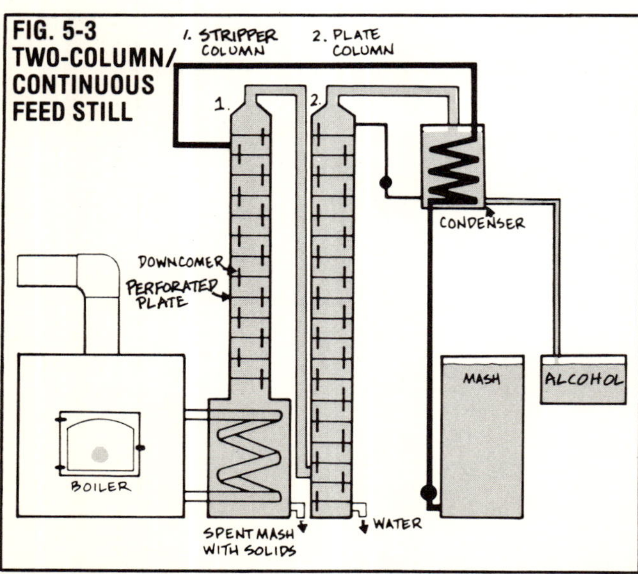


will produce a smaller improvement in the product.

Sounds like a lot of work . . . and it is! But there's an easier way.

## THE PLATE COLUMN

Commercial distillers use a device called a *plate column* which, in effect, performs many distillations during the same run. This is a large (sometimes as much as several *stories* high) column, divided horizontally into a number of sections. (See Fig. 5-2. A *two*-column still is shown in Fig. 5-3.) As the vapors from the still (or *boiler*) rise through the column, they are passed through previously condensed solution at each section. The effect of this is that a new distillation of the rising vapor takes place in each section, so that if the column consists of forty sections, then forty distillations are performed during one run.

While this is a very efficient way to distill alcohol, it isn't practical for the backyard distiller. Besides the immense size, a plate column is very expensive to build and operate. But there are a couple of alternatives that *are* practical for the home distiller.

## THE DOUBLER

A second "proof improving" method is called a "doubler" because—ideally—it doubles the proof of the alcohol during a run without redistilling. In practice, doublers aren't usually quite that efficient . . . however, they do work. (The term "doubling" was used by *moonshiners* in the same way that we've used "redistilling" in this book . . . that is, a rerunning of the removed liquid back through their stills.)

Fig. 5-4 illustrates one common type of doubler which is easy enough to build. In it, the vapor passes up the pipe and is then routed into the doubler, passing through the six inches (or more) of water that it contains. When those vapors have heated the doubler's water temperature to the boiling point of alcohol, the ethanol vapors rise off the surface of the water and pass into the condenser worm . . . which routes them through the condenser where they're cooled into a liquid with a higher alcohol content. How high the percentage will be depends upon the efficiency of the design. The overflow solution from the doubler is then routed through another pipe back into the still where it is revaporized.

The typical "doubler still" often had several doublers hooked up in series . . . forming, in effect, a sort of *horizontal* column (see Fig. 5-5).

The first industrial alcohol made in the United States was produced, in 1801, by a multistage doubling still called Adam's Still. The technique worked then—and it still will—but now there's a *better* way available.

## THE PACKED COLUMN

A packed column is just a simpler, smaller version of a plate column. Instead of a series of complex sections or stages which redistill the vapor, a packed column uses a packing medium such as pall rings (or steel wool) to do the same job. This device isn't as efficient as a full-scale plate column still, of course, but it *is* something that the home distiller can build and operate. (See Fig. 5-6.)

In a packed column, the ascending vapors from the still condense on the surface of the materials which fill the column . . . in fact, those materials are there only to provide a larger surface area for the gases to condense upon. The ascending vapors pass through the *descending* liquids . . . which are the condensed solution (not to be confused with the water in the cooling coil in Fig. 5-6). This system has the effect of redistilling the vapors just as though they were in a plate column (though not so thoroughly), and they become progressively richer in alcohol content. MOTHER's Woodburning Still (see the diagrams on pages 51–52) uses a packed column to distill a higher-grade alcohol from one run than would otherwise be possible.

Of course, after you have distilled your alcohol, much of the liquid *mash* will be left behind, and—depending upon the type of still used—this leftover mixture can either be run through the apparatus again or simply drained off. (In a really *efficient* still, most of the liquid left after one "run" will be water.)

## TESTING FOR PROOF

Earlier we discussed hydrometers and the way they're used to test the alcohol proof of a solution. Proof should be tested at regular intervals throughout your run. Set aside unacceptably weak concentrations, and—when you have a sufficient quantity—run them through the still again.

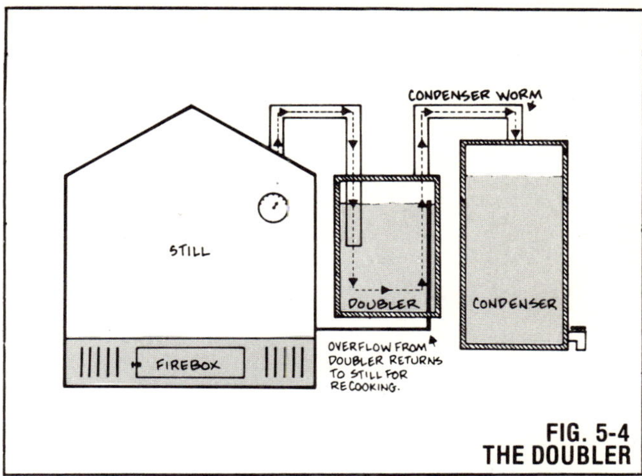


**FIG. 5-4**
**THE DOUBLER**

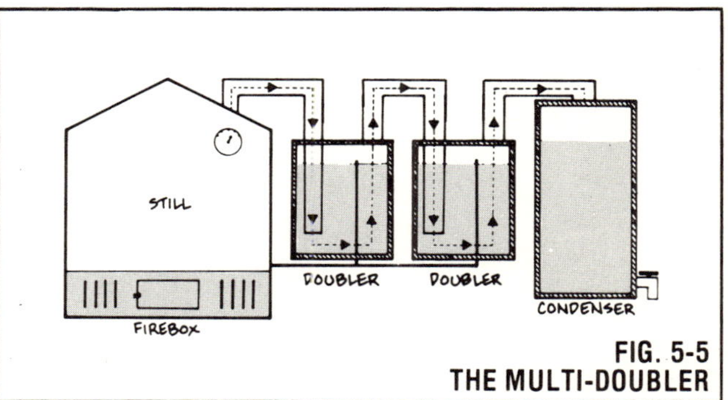


**FIG. 5-5**
**THE MULTI-DOUBLER**

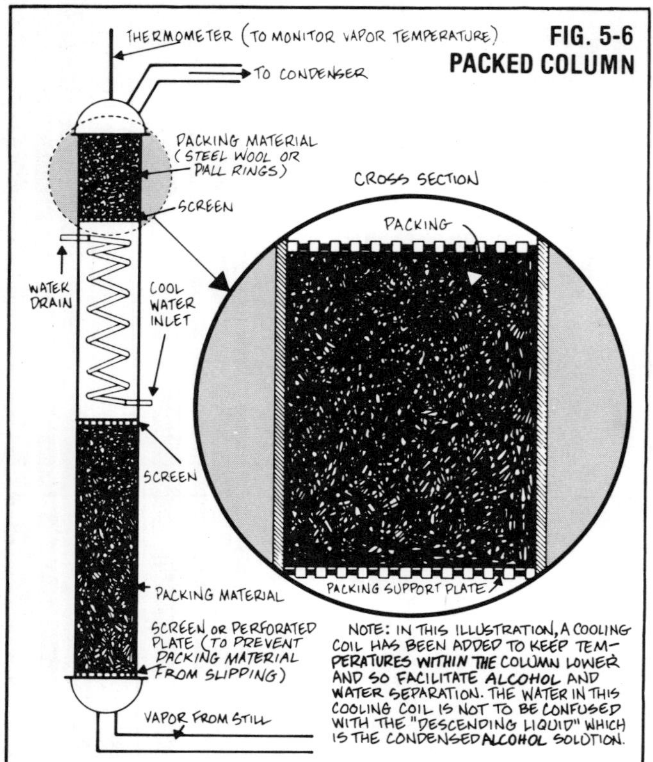


FIG. 5-6
PACKED COLUMN

## DENATURING

The current tax on "drinking" alcohol is about $10.50 per tax gallon, while alcohol for fuel and industrial use isn't taxed (that is, the *manufacture* of it isn't taxed). Because of this, the government is extremely concerned that alcohol produced for the latter purposes not be drinkable. And natural ethanol is technically drinkable (though it might taste awful . . . or actually be poisonous). If you want to produce it legally—either for sale or for your own use—you may have to denature (make unfit for drinking) your ethanol either by using MOTHER's enzymes or by adding a denaturant. The penalties for ignoring this requirement (when it exists) are stiff and rigorously enforced.

In Chapter 8 we'll tell you how to make sure that the alcohol you produce is perfectly legal: what office you need to contact, how you must apply, and the regulations you have to comply with.

# 6 MORE ABOUT STILLS AND THEIR CONSTRUCTION

A basic still (from the word "di*still*ery") is really a pretty uncomplicated device. The still itself is little more than a sealed pot in which provision has been made for the escape of vapors generated in the cooking process. These vapors are routed through a tube (the *worm*) to a cooling vessel called a *condenser*, in which the "steam" condenses into a liquid or *distillate*. The still is filled *partially* with the fermented solution and heated to its boiling point. (Caution: *Over*filling can cause the distillery to "puke" solids into the worm, thus plugging the apparatus and perhaps leading to an explosion.) The higher the concentration of alcohol in the beer, the lower the boiling point of the beer will be. The vapors rising from the ferment are routed to the condenser, and the liquid which results contains a higher concentration of alcohol than did the beer . . . just how *much* higher will, of course, depend on the efficiency of the still.

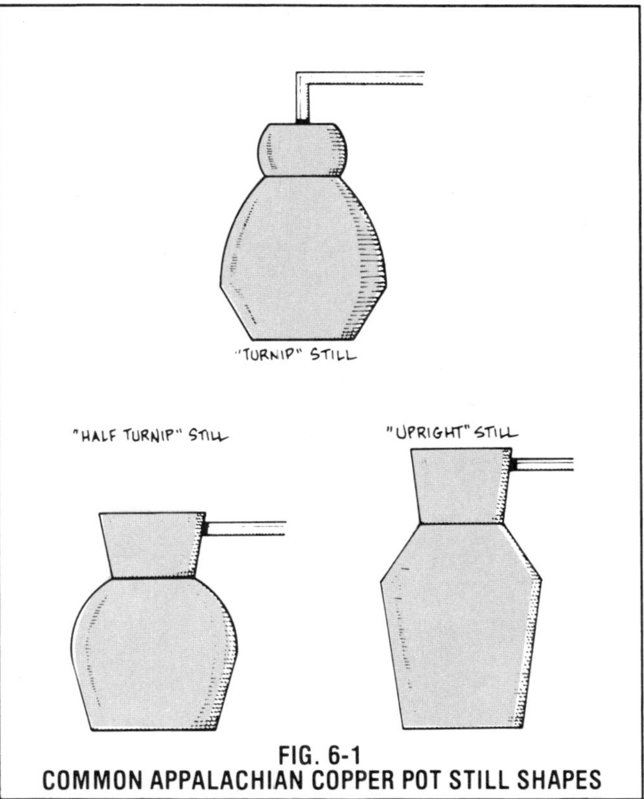

FIG. 6-1
COMMON APPALACHIAN COPPER POT STILL SHAPES

## A BIT OF HISTORY

Good examples of the basic "pot still" used in the manufacture of alcohol are the old-fashioned Appalachian moonshiners' stills (see Fig. 6-1). Designed exclusively for the production of drinking whiskey, such devices were made of copper (to avoid poisonous impurities in the distillate) and were generally small and *always* inefficient. Some whiskey-makers used a very crude system . . . running the liquor from still to condenser to jug. Others—the true corn whiskey artisans—interposed a series of intermediate steps designed to extract poisonous by-products such as fusel oils *and* to increase the proof of the liquor. One such purification step was accomplished in a "thumper" . . . a compartment or group of compartments—between the still and the condenser coil—where the impurities could be trapped because of their tendency to settle out of the lighter alcohol vapor. Thumper boxes were usually filled with fermented mash through which the vapor was passed, entering at the bottom and bubbling up to the top. The deep "thumping" noise which the bubbles made as they percolated through the ferment in the wooden kegs accounts for the device's name.

Simpler versions of the same purification technique—called "slag boxes"—were just small containers, through which the vapor was run . . . the heavier, poisonous oils being periodically drained

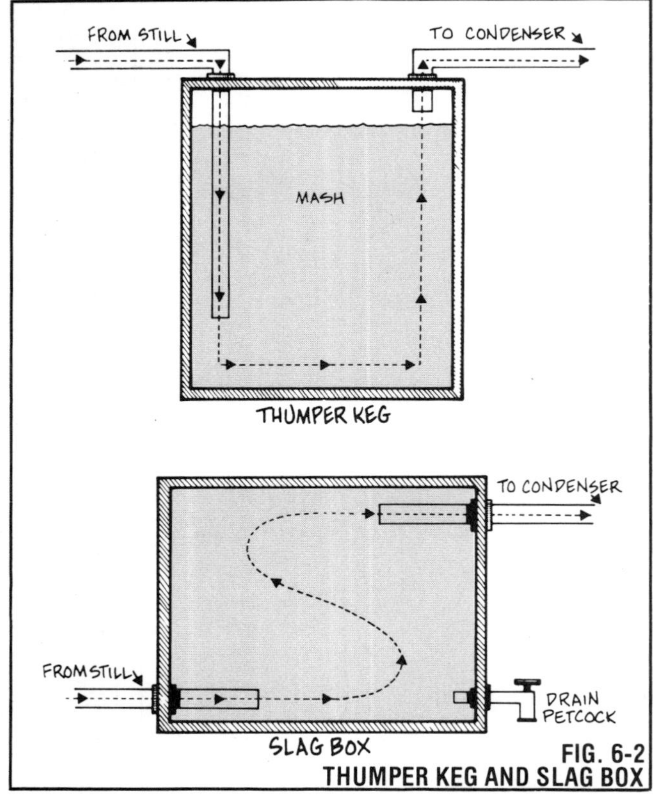


FIG. 6-2
THUMPER KEG AND SLAG BOX

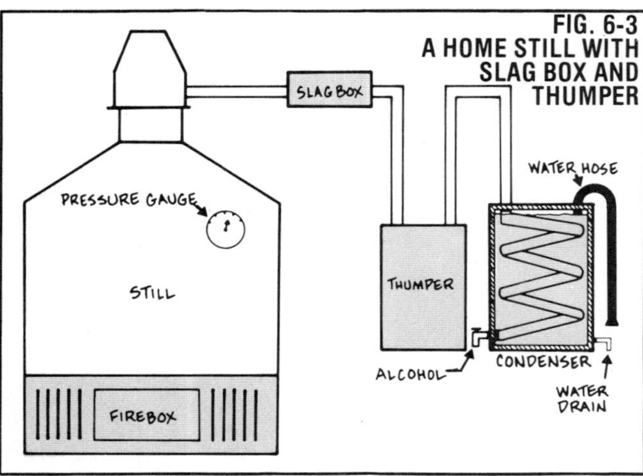


FIG. 6-3
A HOME STILL WITH SLAG BOX AND THUMPER

off. (For illustrations of a thumper and a slag box, see Fig. 6-2 and Fig. 6-3.)

Doublers also served the purpose—described in the previous chapter—of (more or less) "doubling" the alcohol content of the vapor, thus saving the moonshiner from having to make extra runs to produce an acceptably powerful brew.

However—since you're not going to be making ethanol for drinking purposes—you don't have to worry about fusel oils, metal salts, and other contaminants. As long as your still and materials can stand up to heat and moisture without excessive corrosion, you'll have no problems. You don't have to worry about using stainless steel or copper for your still, wood for your fermenting vat, and copper for your worm. Without such concerns, building your own home still for producing alcohol fuel will be much easier and less expensive than the job that faced a prospective moonshiner.

## WHAT'S IN A STILL?

A basic distilling setup for alcohol fuels will consist of a still pot or boiler, a condenser, and a collection tank . . . along with the necessary connecting lines. A more sophisticated and efficient plant will include all of that *plus* a packed or plate column, a proof-testing station, and even a preheater for the beer. In this chapter we'll be giving you plans for MOTHER's Woodburning Still—a unit our researchers designed which will give you about one gallon of *at least* 170-proof ethanol an hour—and some ideas for other (and *larger*) types of stills, as well.* Bear in mind that your choice of still should, of course, depend on the amount of fuel you want to produce. The larger the still, the more ethanol you can make . . . but if your still's too large for your needs, you're just wasting time and money.

You should also consider your available heat sources *and* the grade of alcohol you wish to produce before deciding on a still design. Obviously, you want to run your still on the least expensive, most plentiful fuel you can get. For a lot of people, especially in the eastern United States, that's likely to be wood. Farmers may also have a lot of waste materials—such as corn stalks and wheat straw—which will serve as fuel. Other folks are going to be using everything from coal to solar power to fuel their stills. But remember . . . the point of making ethanol is to supplement or replace other liquid fuels, so it doesn't really make much sense to burn

*See Appendix

heating oil, natural gas, or other scarce resources in the production of your alcohol fuel.

## A "STOVETOP" TEST STILL

There are a couple of reasons why you might want to rig up a simple kitchen still before you go for the full-sized item. First, by making it and running a batch or two through the system, you'll be pacing yourself through a step-by-step (although very elementary) course in the making of alcohol fuel. It'll help give you a "nuts and bolts" understanding of the practicalities involved in brewing ethanol . . . and serve as a preview of what you'll need to do to get seriously involved. Second, it'll give you an inexpensive way to test out your *own* mash recipes . . . without making up large, costly batches that may or may not be effective ethanol producers. True, you won't get as pure an alcohol batch with the test cooker as you will with a packed column still, but running a batch of MOTHER's recipe through—to use as a benchmark against which your own formulas can be measured—will provide a solid foundation of knowledge to work from.

Of course, *before* you cook up a batch of liquid fuel on top of your stove, you have to prepare a small amount of mash. Fig. 6-4 illustrates an easy-to-construct test-batch mash barrel. If you don't want to mess with enzymes and such, when you're running your first test batch just to get the hang of things, here's a simple recipe you can use. Just remember . . . this is a *test mash* to be used for experimentation. It is *not* an efficient or cost-effective recipe for making ethanol.

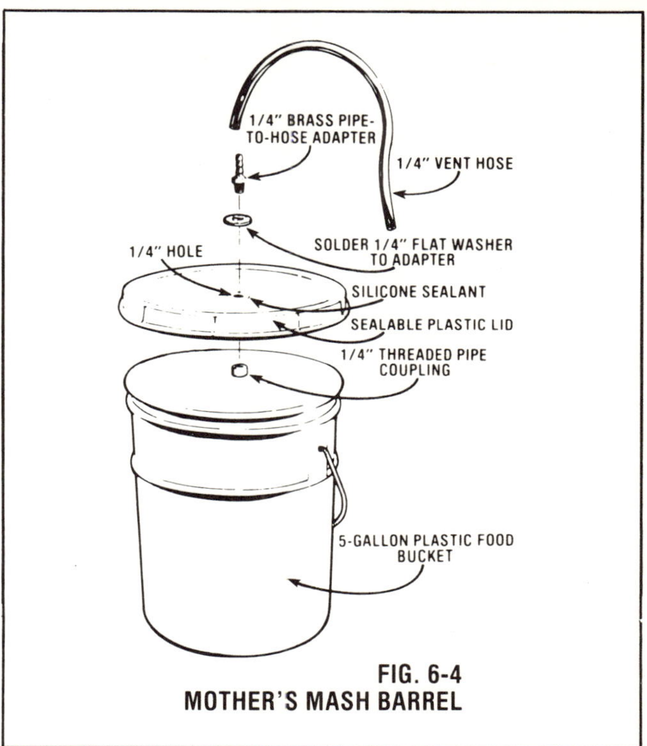


FIG. 6-4
MOTHER'S MASH BARREL

TEST RECIPE

2-3/4 gallons of warm water
1-1/4 pounds cracked corn
2-1/2 pounds sugar
2 ounces yeast
1/2 cup malt

Pour the corn into a porous sack and tie the sack off at the top, leaving room for the corn to expand to half again its volume. Place the bag in warm water and let it sit for a couple of days, giving it a chance to soften (this is the process of liquefaction). Then pour the corn into your mash barrel,

*TOP: You can make a transparent test still from readily available laboratory components. (Mash testing will give you a good preview of large-scale brewing.) ABOVE: This Crock-Pot lab still uses inexpensive kitchen and plumbing hardware, and can be employed to test various mash recipes on a small scale.*

mix in the malt, and add 2-1/2 gallons of warm water. Next, crumble the yeast into a pan containing a quart of water and mix the sugar into it. After about five minutes the mixture will form a thick, foaming liquid and rise. Pour it into the mash barrel on top of your mash, cover it up, and let it do its work. When the fermented mash is ready, pour it into your stovetop test still (Fig. 6-6) and go to work.

Your stovetop still will be simply a slightly more complicated retort still made with readily available materials. (See Fig. 6-5 for a simple retort still.) One of the simplest designs we know of uses an ordinary kitchen pressure cooker for the still (see Fig. 6-6).

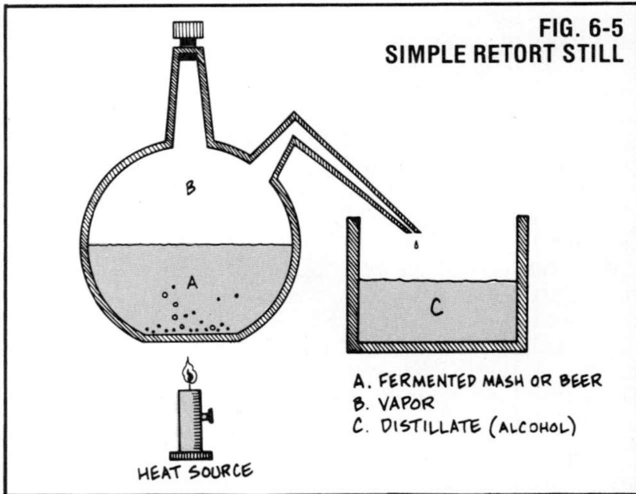


FIG. 6-5
SIMPLE RETORT STILL

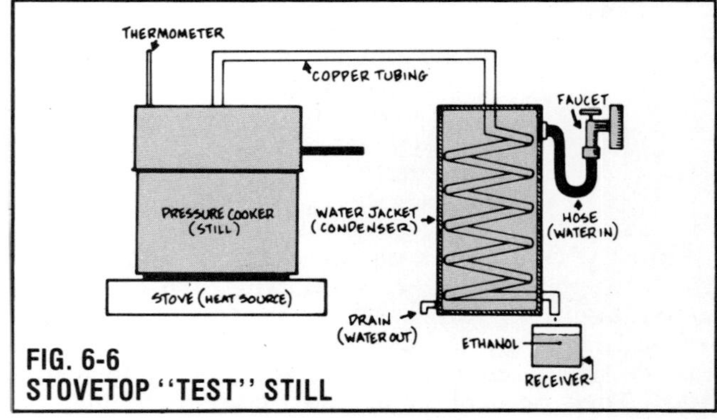


FIG. 6-6
STOVETOP "TEST" STILL

After the fermented mash is poured into the pot, the container is covered and heat is applied. The heat should be controlled carefully so that the temperature of the mash will stay below the boiling point of water (212°F). It is also important to remember that the mash shouldn't fill more than two-thirds of the pot *in any still.* If it does—as mentioned before—the bubbling mash may "puke" into the vapor tube, fouling it and creating a *very dangerous* stoppage which can cause your still—large or small—to explode in your face. Always remember: *Stills are essentially pressure cookers . . . and should be treated with great respect.*

Once the mash in the pot begins to boil, the vapors will rise off the surface, up into the vapor tube, and will then descend into the water-cooled condenser. If a purer grade of alcohol than is initially produced is desired, you can run the liquid through the still again . . . after cleaning the device (especially its tubing) thoroughly. Fig. 6-7 illustrates the best way to alter the pressure cooker lid for use in this test still application.

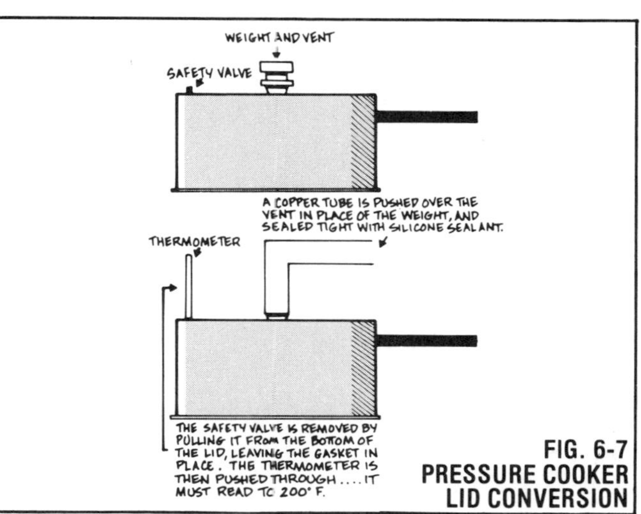


**FIG. 6-7 PRESSURE COOKER LID CONVERSION**

*It's very important to avoid having the alcohol exhaust tubing (or any of the still's internal passageways) become plugged with mash. A pipeline blockage could result in an explosion.*

## MOTHER'S WOODBURNING STILL

This amazing piece of equipment (see Figs. 6-8, 6-9, and 6-10 and Chart 6-1) is nothing more than two identical discarded electric water heater tanks (available at your local dump or in your appliance store's "junk pile"), a few sections of 3" copper pipe, some assorted metal stock, and bits and pieces of plumbing hardware. (Be sure to select tanks that are identical in diameter . . . also, try to get hold of nongalvanized, short, squat containers with a 30- to 50-gallon capacity.)

### CHART 6-1
### BILL OF MATERIALS

| Quantity | Item | Cost |
|---|---|---|
| (2) | 30–50 gallon nongalvanized electric water heater tanks | $ 4.00 |
| 14' | 3/32" X 3" copper conduit | 82.00 |
| 5' | 1/8" X 2" X 2" angle iron | 2.00 |
| 7' | 1/8" X 1/2" X 1" X 1/2" channel iron | 3.00 |
| (1) | 3/16" X 13-1/2" X 13-1/2" steel plate | 1.00 |
| 4' | 3/16" X 1" flat steel | 1.00 |
| 10' | 3/16" X 3" flat steel | 3.50 |
| 40' | 3/8" concrete reinforcing bar (rebar) | 4.40 |
| (1) | 1/8" X 8" X 22" flat plate | 2.00 |
| (1) | 3" X 3" Schedule 40 pipe | .35 |
| (1) | 4" X 6" Schedule 40 pipe | .75 |
| 2" | 1/8" X 1-1/2" X 1-1/2" angle iron | — — |
| (2) | 4" hose clamps | .95 |
| | pall rings | 75.00 |
| 100' | 1/8" copper tubing (1/4" O.D.) | 43.29 |
| 10" | 3/8" (O.D.) copper pipe | .25 |
| (2) | 3/8" copper pipe caps | .20 |
| (3) | 3" couplers | 10.86 |
| (1) | 1-1/2" X 3" X 3" tee | 3.98 |
| (1) | 3" X 3" X 3" tee | 4.18 |
| (1) | 1-1/2" X 3" X 3" X 45° "Y" | 4.09 |
| (1) | 3" 90° elbow | 3.23 |
| 10' | 1/2" rigid copper pipe | 4.56 |
| (4) | 1/2" copper 90° elbows | 1.48 |
| (1) | 1/2" copper 45° elbow | .35 |
| (1) | 1/16" X 1-1/2" X 4" copper pipe | .70 |
| (1) | 1-1/4" to 1-1/2" reducer | .75 |
| (1) | 1/2" to 1-1/2" reducer | .80 |
| (1) | 1/16" X 1" X 1-1/4" copper pipe | .30 |
| (1) | 1/2" X 1" coupler | .45 |
| (1) | 1/2" to 1-1/4" reducer elbow | .45 |
| (1) | 1/2" sweat to 1/2" pipe fitting | .30 |
| (2) | 1/2" to 3/4" bushing | .35 |
| (1) | 1/2" X 3" thermometer well | 1.05 |
| (2) | 0°F–212°F straight thermometers | 8.25 |
| (1) | 0°F–250°F threaded thermometer | 6.48 |
| (1) | 3/4" X 2" nipple | .55 |
| (1) | 3/4" gate shutoff valve | 3.10 |
| (1) | 3/8" drain cock | .95 |
| (1) | 1/16" X 3/4" X 3" copper pipe | .50 |
| (1) | 1/16" X 6" X 10" copper plate | 1.65 |
| (2) | 1/4" X 4" eyebolts | .60 |
| (4) | 1/4" flat washers | .10 |
| (2) | 3/8" X 2" compression springs | .60 |
| (2) | 1/4" wing nuts | .10 |
| (6) | 3/8" hex nuts | .12 |
| (3) | 3/8" X 1" machine bolts | .15 |
| (1) | 3/8" X 2" machine bolt | .05 |
| (2) | 1/4" X 1-1/2" machine bolts w/nuts | .15 |
| (1) | 1/2" X 1" compression spring | .20 |
| (1) | 3/8" flat washer | .02 |
| (1) | 1/8" X 1" X 6" pipe section | .25 |
| (2) | 1/8" tubing threaded unions (if desired) | .30 |
| (6) | 1/8" sweat to 1/4" pipe fittings | .90 |
| (6) | 1/4" pipe to 1/4" hose barbs | 1.05 |
| (2) | 1/4" brass tees | .65 |
| (2) | 1/4" male to female brass elbows | .70 |
| (3) | 1/4" needle valves | 6.42 |
| (1) | 1/2" brass tee | .45 |
| (4) | 1/4" to 1/2" bushings | 1.00 |
| (2) | 1/4" close nipples | .30 |
| (1) | 1/2" sweat to 1/2" pipe fitting | .24 |
| (2) | garden hose to 1/2" pipe adapters | .85 |
| (1) | garden hose "Y" adapter | .98 |
| 12" | baling wire | — — |
| 3' | 1/4" air hose (cut to necessary length) | 1.48 |
| (1) | 4" stovepipe 90° elbow | 2.09 |
| (1) | length 4" stovepipe (as needed) | — — |
| 1 piece | fiberglass insulation batting (if desired) | — — |
| | Total Cost | $302.80 |

NOTE: The figures above represent new material prices as of 7/1/79. By scrounging and buying from salvage or scrap dealers (especially with regard to the copper items), the total cash outlay for your woodburning still project can easily be halved.

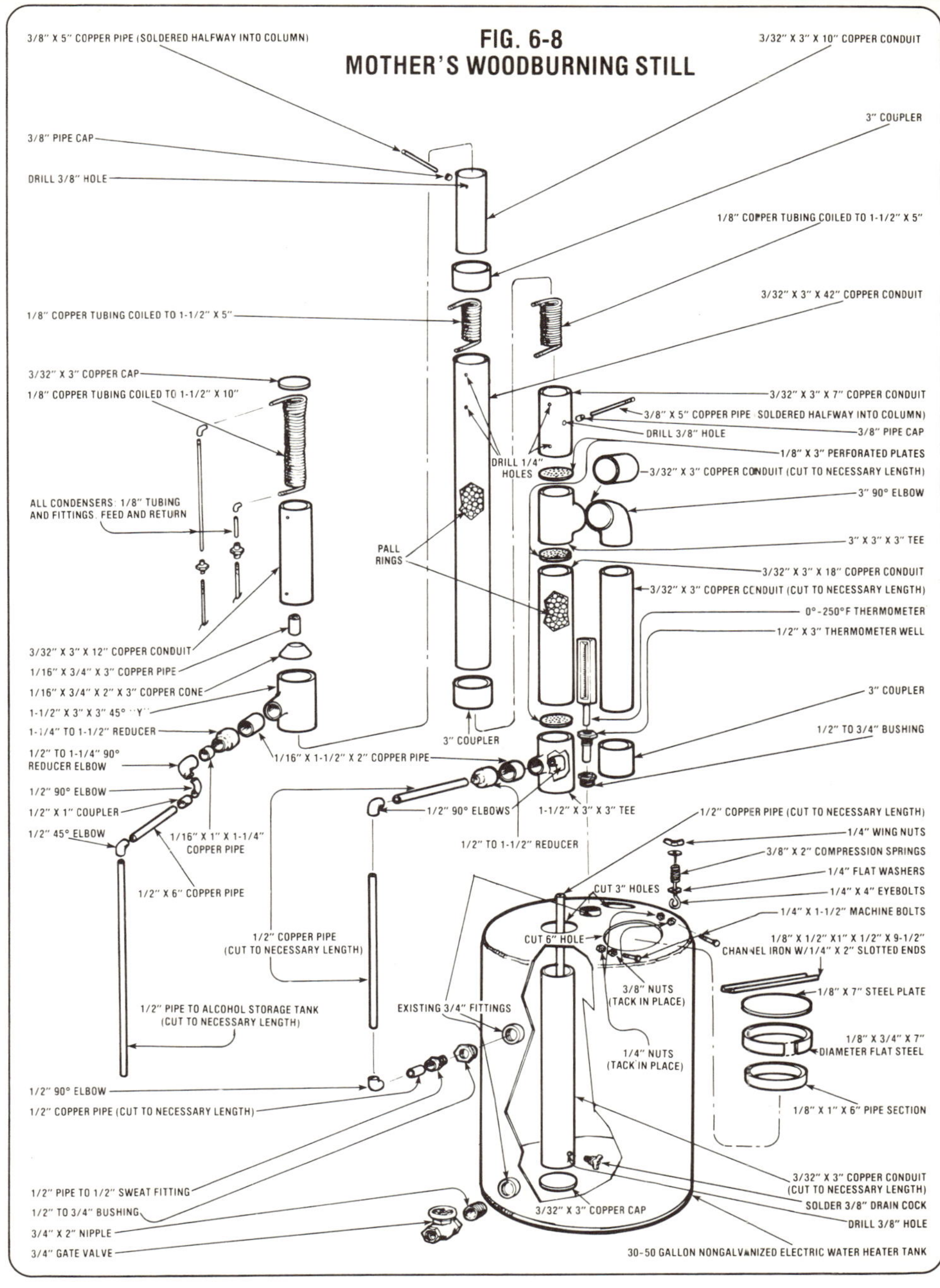
FIG. 6-8
MOTHER'S WOODBURNING STILL
3/8" X 5" COPPER PIPE (SOLDERED HALFWAY INTO COLUMN)
3/32" X 3" X 10" COPPER CONDUIT
3/8" PIPE CAP
DRILL 3/8" HOLE
3" COUPLER
1/8" COPPER TUBING COILED TO 1-1/2" X 5"
1/8" COPPER TUBING COILED TO 1-1/2" X 5"
3/32" X 3" X 42" COPPER CONDUIT
3/32" X 3" COPPER CAP
1/8" COPPER TUBING COILED TO 1-1/2" X 10"
3/32" X 3" X 7" COPPER CONDUIT
3/8" X 5" COPPER PIPE SOLDERED HALFWAY INTO COLUMN)
DRILL 3/8" HOLE
3/8" PIPE CAP
1/8" X 3" PERFORATED PLATES
DRILL 1/4" HOLES
3/32" X 3" COPPER CONDUIT (CUT TO NECESSARY LENGTH)
3" 90° ELBOW
ALL CONDENSERS: 1/8" TUBING AND FITTINGS, FEED AND RETURN
PALL RINGS
3" X 3" X 3" TEE
3/32" X 3" X 18" COPPER CONDUIT
3/32" X 3" COPPER CONDUIT (CUT TO NECESSARY LENGTH)
0°-250°F THERMOMETER
1/2" X 3" THERMOMETER WELL
3/32" X 3" X 12" COPPER CONDUIT
1/16" X 3/4" X 3" COPPER PIPE
1/16" X 3/4" X 2" X 3" COPPER CONE
1-1/2" X 3" X 3" 45° "Y"
3" COUPLER
1-1/4" TO 1-1/2" REDUCER
3" COUPLER
1/2" TO 3/4" BUSHING
1/2" TO 1-1/4" 90° REDUCER ELBOW
1/16" X 1-1/2" X 2" COPPER PIPE
1/2" 90° ELBOW
1/2" X 1" COUPLER
1/2" 90° ELBOWS
1-1/2" X 3" X 3" TEE
1/2" 45° ELBOW
1/16" X 1" X 1-1/4" COPPER PIPE
1/2" TO 1-1/2" REDUCER
1/2" COPPER PIPE (CUT TO NECESSARY LENGTH)
1/4" WING NUTS
3/8" X 2" COMPRESSION SPRINGS
1/4" FLAT WASHERS
1/2" X 6" COPPER PIPE
CUT 3" HOLES
1/4" X 4" EYEBOLTS
1/4" X 1-1/2" MACHINE BOLTS
CUT 6" HOLE
1/8" X 1/2" X1" X 1/2" X 9-1/2" CHANNEL IRON W/1/4" X 2" SLOTTED ENDS
1/2" COPPER PIPE (CUT TO NECESSARY LENGTH)
3/8" NUTS (TACK IN PLACE)
1/8" X 7" STEEL PLATE
EXISTING 3/4" FITTINGS
1/2" PIPE TO ALCOHOL STORAGE TANK (CUT TO NECESSARY LENGTH)
1/4" NUTS (TACK IN PLACE)
1/8" X 3/4" X 7" DIAMETER FLAT STEEL
1/2" 90° ELBOW
1/2" COPPER PIPE (CUT TO NECESSARY LENGTH)
1/8" X 1" X 6" PIPE SECTION
3/32" X 3" COPPER CONDUIT (CUT TO NECESSARY LENGTH)
1/2" PIPE TO 1/2" SWEAT FITTING
SOLDER 3/8" DRAIN COCK
1/2" TO 3/4" BUSHING
3/32" X 3" COPPER CAP
DRILL 3/8" HOLE
3/4" X 2" NIPPLE
3/4" GATE VALVE
30-50 GALLON NONGALVANIZED ELECTRIC WATER HEATER TANK

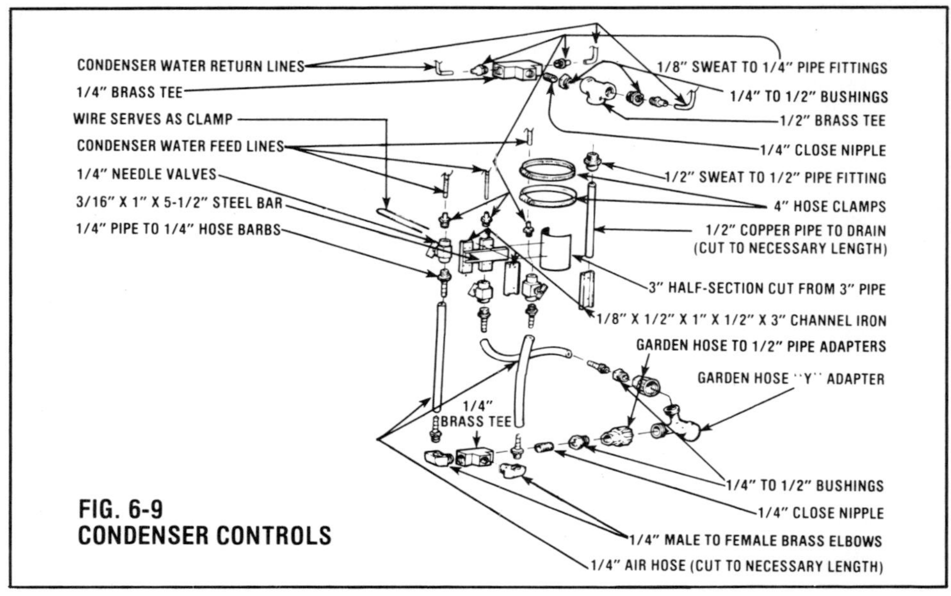


FIG. 6-9
CONDENSER CONTROLS

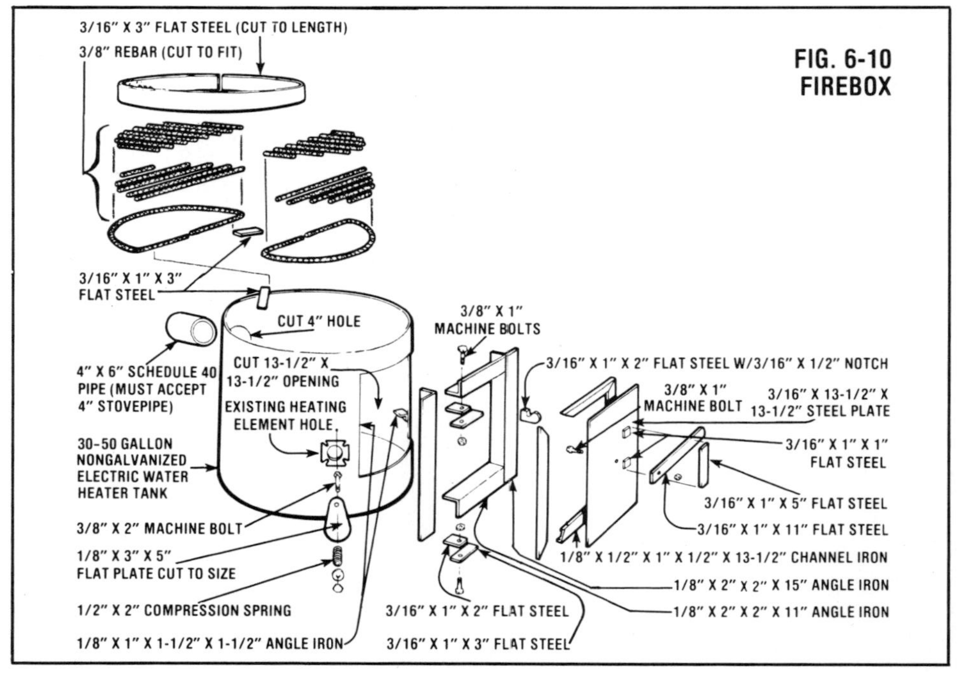


FIG. 6-10
FIREBOX

## How It Works

The mash solution is contained in a hot water tank and is constantly simmered by a blaze set in a firebox underneath the container. The vapors that rise from the steaming liquid travel up through a short length of 3″ conduit and into a five-foot section of pipe . . . which is packed with pall rings and contains a 5″ length of coiled copper tubing at each end of the pipe.

As the alcohol/water "steam" passes through its "maze" of pall rings, most of the water will be left behind on the surface of the ceramic "U-shaped" pieces (the pall rings provide a far greater surface area than would the inside of the "bare" pipe). In addition—by running cool water through the coiled tubes at a controlled rate—the condensation process can be regulated to provide the most efficient separation of alcohol from water.

Ideally, only pure alcohol vapors will rise to the top of the conduit, while the water should fall to the bottom of the sealed tube. In practice, however, some $H_2O$ *does* find its way past the pall rings in the form of steam, and—by the same token—much of the "pure" water that falls back into the small reservoir is actually laden with a good deal of recoverable alcohol. Therefore, an additional plumbing circuit has been incorporated into the base of the copper tower . . . to allow some of this valuable alcohol/water mixture to flow *back* into the mash solution and be recycled.

At the top of the four-foot column—just above the upper series of coils—is an inverted funnel which is surrounded by still another condenser. When the "stripped" alcohol vapors pass through this cone and make contact with the coils, the ethanol liquefies, runs past the outer surface of the funnel, flows through a downspout, and enters a storage vat.

## Let's Build!

Before you start "still-building", you'll need to gather an oxyacetylene torch (with welding and cutting tips), a pipe-cutting tool, several C-clamps or vise grips, a hammer, a ruler, and a file. Begin by cutting around the circumference of one of your salvaged water tanks at a height of about 18 inches. Then fashion an opening in the side of the container to accommodate the firebox door. Next, weld together an angle-iron frame for this opening as shown, fasten it to the tank, and cut a steel plate to

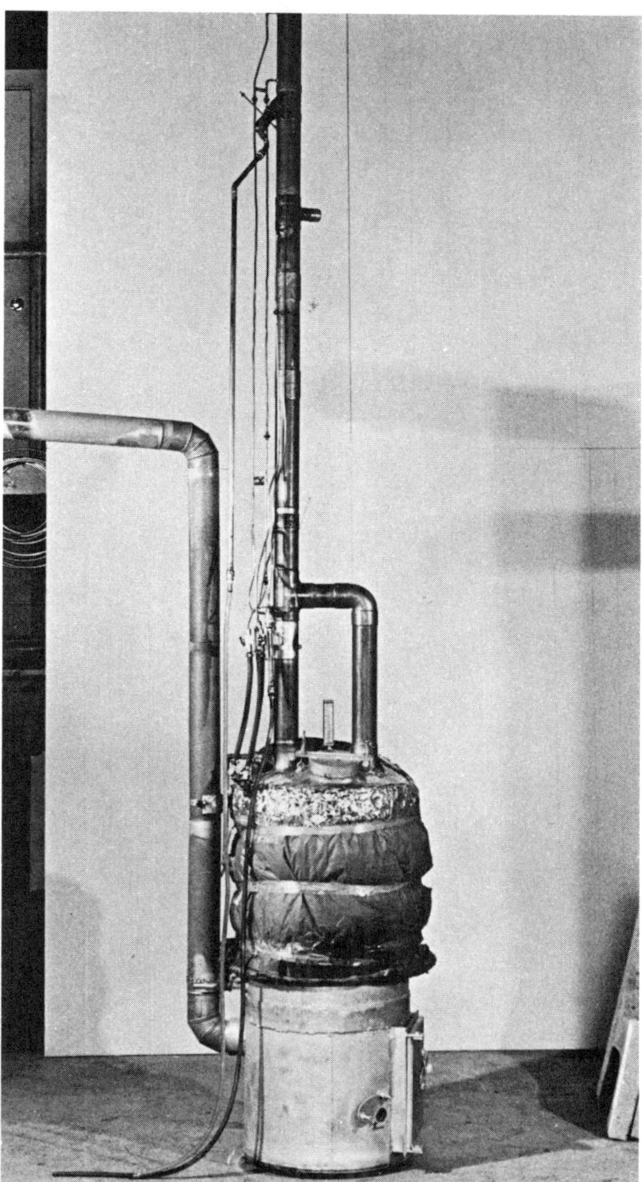

*This compact still has a production capacity of about 3/4 gallon of fuel-grade ethanol per hour, can be made from mostly scrap materials, and could cost as little as $150 to build.*

*TOP: The still's mash vat should have a drain valve at its lowest point to release the spent liquid after the alcohol has been driven off. This hot water can be reused for the next mash batch. ABOVE: The firebox shown here is designed to burn wood.*

serve as your cooker's "portal". Border the inner surface of this lid by welding four sections of trimmed-to-size channel iron in place, then go on to fabricate your two hinges and the latch system as illustrated . . . and fasten these parts in position.

With this done, weld a bolt—head first—to your tank (at a point just above an existing heater element hole), then make a small "draft door" from a piece of steel plate and attach it to this stud . . . using a spring, washer, and nut so the plate can pivot. Cut a hole in the rear of your new firebox and weld the 4" stovepipe collar to the circumference of this opening . . . then fabricate a grate—as shown—from 3/8" rebar, and tack its supporting tabs in place. (Form the rod "frame" for each grill half to fit the inside of the tank, and make a weld at the junction point of both grate sections. Then tack the rearmost platform directly to the tank. The front grill will rest solely on the tabs which you've welded to this fixed grate *and* to the inner face of the firebox.)

Finally, take your 3/16" X 3" length of flat stock and weld it to the outside of the water container to serve as a girth band. (Since this midriff support is actually the only joint between the firebox and your mash vat, it must be securely fastened to both containers. You can assure a tight seal by welding one end of the iron strap to the fire chamber and "guiding" the band around the vessel's circumference with torch and hammer . . . tack-welding as you go. When the circle is completed, force the remaining water tank into the iron belt border atop the firebox section, and run a continuous bead along both upper *and* lower edges of the metal band.)

Now you're ready to start construction on the still itself. First, cut two holes—in the tank's top—large enough to fit the 3" brass tee and the coupler, then slice a third access hole into the "dome" of the container as shown. Fashion a sealable lid for this opening by welding a collar around the hole, attaching a slightly larger ring to a piece of circular flat plate, and making a "dog and bar" latch as depicted in the illustration. (You may also want to line the edge of the cap with a liberal amount of silicone sealant . . . to prevent any vapor loss during the distillation process.) Finally, *braze* the tee fitting and coupler to the tank to assure a solid joint.

For the next step, cut one length each of 3" and 1/2" conduit, both long enough to reach nearly from the bottom of the tee to the base of the container, then seal the larger pipe's lower end with a cap and install a drain cock as illustrated. Solder this entire assembly to the lower collar of the tee . . . then take the length of 1/2″ pipe and fasten a 90° elbow to one end of it. With this done, solder the short 1-1/2″ nipple and reducer to the tee . . . then slip a length of 1/2″ copper tube through the reducer's opening and solder it to your elbow-and-pipe assembly as shown. Now just route additional tubing to an available existing fitting on the water tank to complete a circuit from the inside of the small "reservoir" to the main mash storage vat.

You can now install a 3/4″ gate valve in the base of your container and a thermometer at its top. Most tanks will already have threaded fittings in convenient locations, so it's just a matter of installing adapters—if necessary—and attaching the hardware. At this time, too, check over both the firebox and the vat for extra openings and plug them.

From this point, you can complete the remainder of the tower as follows: Trim each section of 3″ conduit to the length indicated. Then make three perforated disks . . . cut to fit snugly inside the tee fittings (they should rest against a lip inside the three-way unions) and each drilled with a series of 1/4″ holes about 1/8 inch apart. Place one of your disks into the top opening of the tee, and solder the 18″ length of pipe in position.

Next, fill this section of conduit with pall rings, put your second perforated disk into the 3″ X 3″ X 3″ tee, and solder that fitting in place. (At this point you can also permanently fasten the pipe connection between the coupler brazed to the top of the mash container and the horizontal arm on the brass tee.)

Now simply install your final drilled plate at the top of the tee, and stack the remaining components in the order shown. Don't forget to install the thermometer wells in the two sections of conduit as illustrated (these cylinders must protrude *into* the pipe, but have to be capped on their "inside" ends to prevent vapor leakage). Also, remember to fill the long section of 3″ conduit with pall rings before sealing it up.

The condenser coils can be made by winding copper tubing—tightly—around a piece of 1-1/2″ pipe until you produce a "spring" of the desired length. (Remember to solder-seal the joints between the column and the protruding copper "pipelets".)

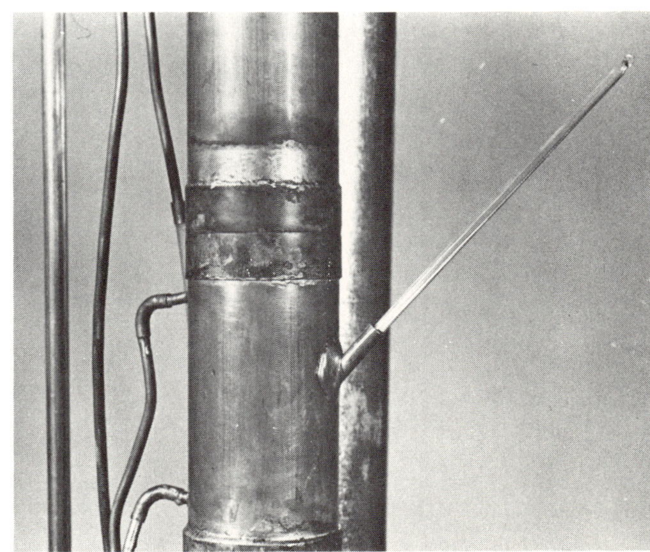

*TOP: The main tank thermometer—which is calibrated in degrees centigrade (Celsius)—indicates the temperature of the vapor in the vat. ABOVE: Another thermometer is located in the column for the purpose of monitoring condenser adjustment.*

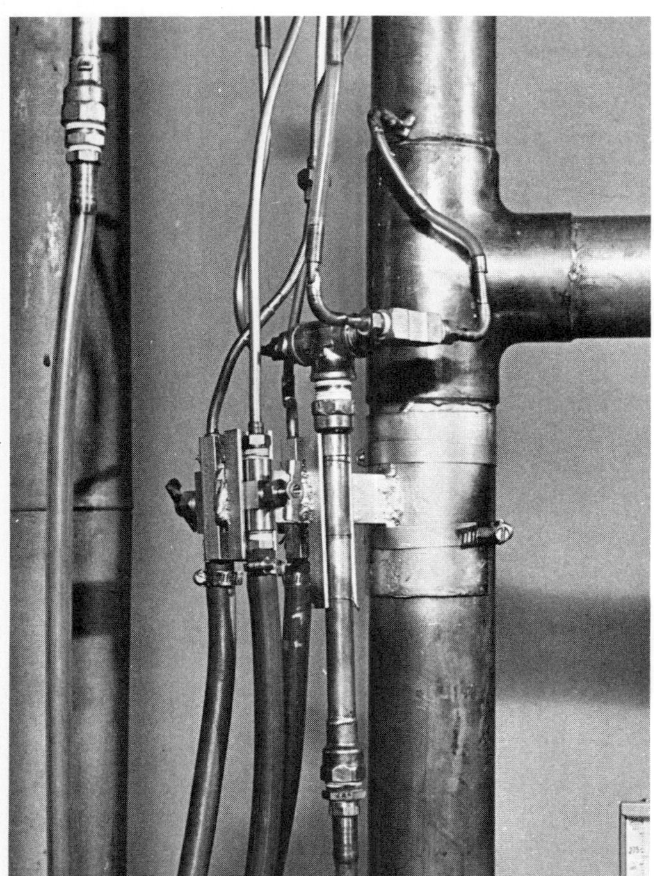

*Three needle valves control the flow of cooling water to the condensers, while the clear tube—shown at left—delivers a slow but steady stream of 85% pure alcohol to a holding tank.*

Unless you have a small funnel that fits snugly within the 45° "Y" at the tower's top, you'll have to make one . . . but this involves nothing more than forming a sheet of copper (or a section of copper pipe) into a cone shape, soldering the seam, and fastening a 3/4" pipe to its "small" end. Be certain the funnel is leak-free—*and* carefully soldered to the "Y"—or some of your distilled ethanol will leak back into the mash vat.

Now complete the copper plumbing circuit to your alcohol storage tank, and hook up the three pairs of water feed and discharge lines as indicated. Be certain that each feed line is equipped with a needle valve, since careful water control is essential to the distillation process.

## You're Ready to "Make a Run"

First, wrap the "boiler" tank with fiberglass insulation batting, and then begin the distillation process by filling your mash container with the fermented solution to a level several inches below the lip of the access hole (this might total 25 gallons or more, depending on how large your water tank is). Next, connect your stovepipe to an outside chimney, "lay a blaze" in the firebox, and wait for your mash to generate some alcohol/water "steam". (This will occur when the tank thermometer registers in the vicinity of 170°F. Remember that this "degree sensor" is positioned to indicate *steam vapor*—rather than liquid—temperature.)

Now open the three condenser flow valves, and wait for the vapors to reach the 175°F to 190°F range. At this point the still's column will begin to hiss . . . and you can start "fine tuning" the still's controls. The main condenser coil at the top of the column doesn't need continual adjustment . . . just let water flow through it at a slow, steady rate. The other two coils, however, must be controlled carefully. During the first two hours or so of distillation, the temperature at the lower primary coil should be maintained at a point between 176 and 182°F, while the heat in the vicinity of the upper condenser should be held to about 170°F (to cool either area down, just *increase* the flow of water through the coils).

When everything is adjusted properly, an 85%-pure alcohol product will begin to flow into your storage container. The stream should be slow but steady . . . if it's excessive, chances are that you're getting too much water in your alcohol. (You can prevent this undesirable situation by cooling your upper and lower primary coils slightly . . . but *do* try to stay within the temperature ranges previously indicated.)

Be certain to check your fire regularly also. During most of the run, you'll discover that the vapor temperatures in the mash vat will remain around 180°F . . . but, as more alcohol is driven from the solution, that figure will rise. When the heat in the vat reaches about 200°F (after perhaps three hours) the condensed product will be mostly water . . . and it's best to shut the apparatus down, since you'll just be diluting your alcohol.

It'll take a few "runs" to get the feel of how to control your still for best results. The "tail end" product will be fairly weak . . . but the first two-thirds of the run should yield good 170-proof (*and* higher) ethanol. MOTHER's still produced about two gallons of alcohol from a total mash solution of 30 gallons. You can expect a flow rate of about 3/4 gallon an hour with excellent results . . . if you try to hurry the process, however, be ready to settle for a weaker ethanol solution.

Be aware, too, that you can run your distilled product through the apparatus for a second time. This procedure will drive most of the excess water out of the solution and produce nearly pure (190-proof) alcohol. You may also want to save the "tail end" of each run and mix it with your next mash solution. Since such "dregs" contain 60- to 100-proof alcohol, they will increase the strength of the following batch of fuel.

Finally, remember that—since this distillation apparatus is capable of producing a sizable amount of alcohol during the course of each "run"—Federal Bureau of Alcohol, Tobacco, and Firearms regulations must be strictly adhered to. When you get to the point where you are producing an alcohol product on a regular basis, be sure to contact your local ATF regional office for precise instructions on how to handle, store, and provide security for your homemade ethanol.

## "STILL" MORE

You probably won't want to build a unit that's very much *smaller* than MOTHER's woodburning steamer if you're serious about making alcohol fuel for your own use or for sale. But you might want to build one that's larger. If so, you're going to need to figure out what capacity you *do* need, and then

*MOTHER's six-inch packed-column still holds about 250 gallons of mash and can distill 6–8 gallons of 160- to 180-proof ethanol fuel per hour. Most of its parts are available as scrap.*

*TOP: A perforated disk welded within the column near the flange joint holds the packing material (in MOTHER's 6" column still) in place. ABOVE: The tightly sealed or gasketed access door is held down with eyebolt "dogs", springs, and washers.*

design a system which is appropriate for that goal.

If this seems too complicated or puts you off, slow down. It's not really that difficult. Remember the stills we've already described for you in this chapter. MOTHER's woodburner is nothing more than a sophisticated, high-efficiency version of the simple little pressure cooker stovetop unit we showed you how to make. And a larger still—one that can pump out a couple of hundred gallons per run—is just the "big brother" to MOTHER's baby.

## MOTHER'S SIX-INCH COLUMN STILL

The correct designations for MOTHER's still column designs are *packed* or *differential columns*. A packed column—as mentioned before—is one that is stuffed with some sort of material that provides surface for vapor contact between the phases of fractionating. Three different types of packed columns are used in today's industry:

[1] Those that use a conventional packing—such as ring packings and saddle packings—which is dumped or random-packed into the column.

[2] Those that use a systematic and geometrical packing, with the packing units placed by hand in particular reference to each other.

[3] A "pseudoplate" column where plates of various designs are used.

MOTHER's six-inch column is of the conventional random-packed type. The column is filled with 5/8-inch pall rings. They provide approximately 131 square feet of surface per cubic foot and at the same time allow about 90% free gas space.

The packed column is nothing new as far as still designs go. The differences between our design and the "traditional" models occur when introducing the vapors into the column and removing the heat from the column. The usual method is to mount the column on the batch pot and feed the vapor into the tube at the very bottom. Since the column must maintain temperature equilibrium (that is, a decreasing temperature rate in the column, starting with the boiling temperature of the mash at the bottom and ending with approximately 175°F at the top), this method controls that equilibrium by regulating the amount of heat in the vapor as it is introduced into the column.

Both of MOTHER's stills increase their distillate production rate by forcing more alcohol vapor into their columns than a normal unit could handle. Initially, the concept sounds paradoxical. (If *too* much vapor is introduced into the column, the column becomes too hot and therefore should reduce the effect of the still to the equivalent of a simple distillation process.) To overcome this problem, two heat exchangers are built into the column, which simultaneously extract heat and water while leaving the alcohol vapors. These heat exchangers bring the column into equilibrium and increase the output of distillate from six gallons per hour to eight gallons per hour. Of course, nothing is free in the energy business, and the price that has to be paid for the increased alcohol output is the use of extra amounts of cooling water.

There are two ways of building a "reprocessing-type" still. With the reboiler approach (Fig. 6-11)

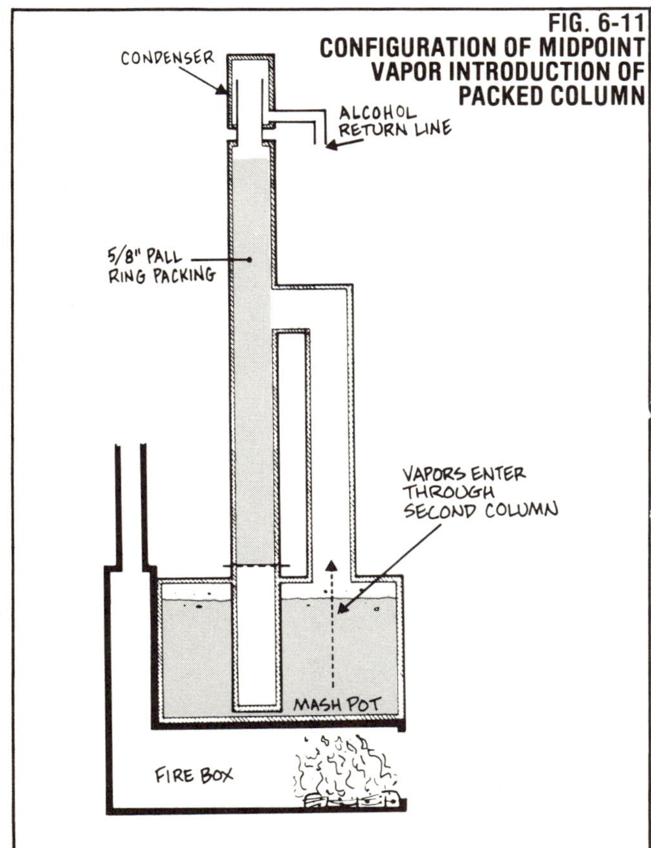


FIG. 6-11
CONFIGURATION OF MIDPOINT VAPOR INTRODUCTION OF PACKED COLUMN

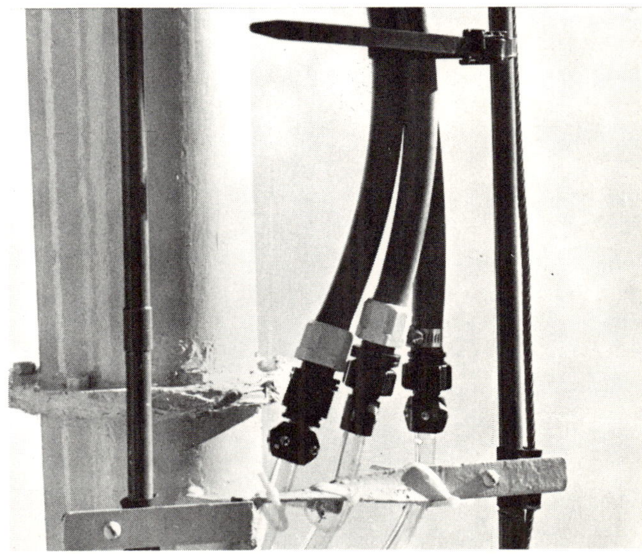

*TOP: The water inlet for the two heat exchangers and the condenser coil is trisected by means of plumbing fittings, and flow is controlled with separate needle valves. ABOVE: The still's transparent tubes allow a visual check of the water flow.*

*To allow for easy transportation of the 6" still, the entire column was built with a hinged bottom. A ring of bolts and nuts serves to fasten each of the column sections firmly to the cooking tank when a batch of mash is run through the still.*

the vapor is channeled up a separate column and introduced at the midpoint of the packed column. Immediately, the vapor hits the cooling coil (heat exchanger) and its temperature is reduced to 185 °F. The vapors are condensed to a liquid, and the heat rising from the reboiler revaporizes them. Because of the partial vapor pressure phenomenon, more alcohol is revaporized than water, so the alcohol ascends in the column and the water descends. This vapor/liquid transfer will reoccur, time and again, throughout the entire length of the column (if the equilibrium is maintained). Any alcohol that enters the reboiler will "reboil" back up into the column.

The second method (Fig. 6-12) does away with the

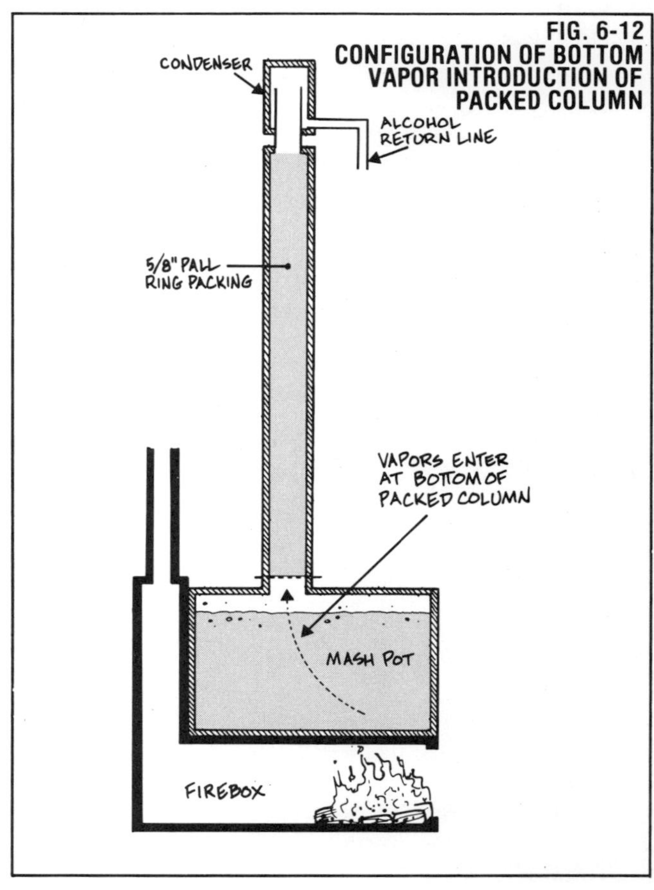


FIG. 6-12
CONFIGURATION OF BOTTOM VAPOR INTRODUCTION OF PACKED COLUMN

reboiler *and* the second column, and introduces the vapor into the column at its bottom . . . but here, again, there are two heat exchangers in the column. Not enough testing has been done yet to show definitive advantages or disadvantages of either of MOTHER's designs. One *positive* feature of the reboiler design is that the water stripped from the alcohol is pure and clean. And, since the water is already hot, it can be used for cooking the next batch of mash. However, if you do not expect to be running a continuous operation, the "bottom vapor introduction" design is probably the best method of building a still, since the manufacturing costs are lower.

In both methods, reliable thermometers must be placed within inches of the top side of the heat exchangers in the column. Depending upon the sophistication desired, this column can be controlled manually or with instrumentation.

The size of the still can be scaled up or down: The principle will work regardless of the size. A larger tank for the batch pot can be used, and the column can be increased to eight or ten inches (you can use a ratio of about 24:1, height to diameter).

The firebox need not be all metal like MOTHER's (it was made this way to make the still movable), but can instead be made of stone and brick.

In all, this is a very flexible still, capable of accepting many modifications. You can even obtain as high as 190-proof alcohol by returning some of the distillate from the top condenser for reflux. However, for an alcohol fuel, all that is needed is 170 to 175 proof . . . and this still will give you this moderate proof at a high gallon-per-hour delivery.

The point is that it doesn't matter much whether you're making two gallons or two thousand gallons of liquid fuel . . . a still is a still. To design your own setup you need to include the following:

[1] **Mixing**, **cooking**, and **fermentation tanks**, equipped for agitating the mash. (Many large systems use the same vat for mixing, cooking, and fermentation . . . and since some successful farm-based units use *5,000-gallon* tanks, you can see why duplication of materials is avoided!)

[2] A **heat source** for cooking and for warming the fermentation tank, as well as for vaporizing the alcohol in the fermented mash.

[3] A sophisticated **"doubling" system** . . . that

is, a packed column or (in most large systems) a plate column.

[4] A **condensing tank**.

[5] **Pumps** to move materials from step to step of the process, and the plumbing for same.

[6] A **holding tank** for the fermented mash.

[7] A **storage** and **denaturing tank** for the finished ethanol.

In a farm-scale alcohol system, you'll also want to include provisions for filtering the beer and separating the valuable solids (DDGS).

Fig. 6-13 shows the design for a basic farm-based distillery put together by researchers at Iowa State University. Don't let the maze scare you off . . . it's not as complicated as it appears to be. Here's the way they describe it:

## *SYSTEM DESCRIPTION*

*A boiler is stoked with corn stalks developing steam at a pressure of 15 PSI. The steam is injected into a large mild steel tank holding approximately 1,650 gallons of a mixture 60–75% water and 25–40% mash corn. The tank will have a conical bottom to aid in the removal of the leftover*

*TOP: A cooker/fermentation vat. LEFT: An alcohol production system—based on MOTHER's oil furnace—which consists of the furnace, the cooker/fermentor, and a 6" packed-column still. ABOVE: The hot-oil furnace is set up for woodburning.*

*corn mash after fermentation. It will also have a bulk tank stirrer to agitate the mixture so uniform cooking will be assured. The steam will be injected into the bottom of the tank and will cook the mixture as it rises.*

*After the fermentation is over, the mash settles out and the 6.5% beer can be removed. After it leaves the cooker-fermenter, the beer passes through a screen, removing any large particles*

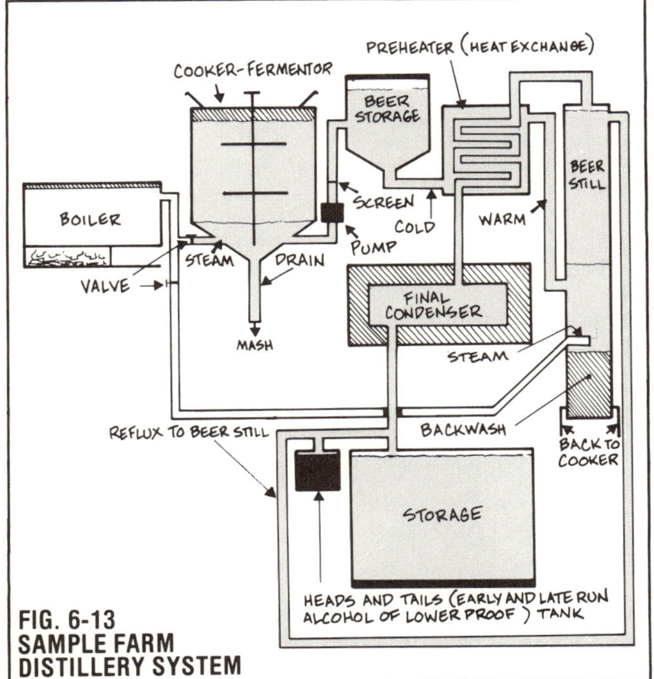


**FIG. 6-13 SAMPLE FARM DISTILLERY SYSTEM**

*On the left is the main column of the still . . . on the right is the reflux. Note the condenser at the junction of the two columns.*

This double network of pipes and pumps moves the wood-heated oil from the furnace to the cooker or the still, and transports the liquid mash from the cooker to the main tank of the still.

*that might have been pumped out of the tank. From here it enters a storage tank which holds it until the beer still is ready. This tank can be a little smaller than the previous tank because the mash has been removed. It will also be conically shaped on the bottom end for easy removal of its contents.*

*From here the beer goes into a heat exchanger which uses copper coils filled with evaporated ethanol from the column to preheat the beer before it goes into the still. As the flow comes into the column, it is hit with steam rising from the bottom of the still where it has been injected. The beer still consists of a column packed with either fiberglass or clay chips. As the steam comes up and hits the liquid beer, it evaporates the ethanol, giving a high percent of alcohol. As the alcohol is coming out of the top of the still, it passes through the heat exchanger previously mentioned and then into a final condenser. From there it is either refluxed back to the still or is put back into a storage tank.*

## BOILER CALCULATIONS

### Requirements

*Steam requirements were based on figures received from the following companies:*

*ICFAR—Indianapolis Center for Advanced Research, Indianapolis, Indiana.*

*ACR—ACR Process, Urbana, Illinois.*

*GP—Grain Processing, Muscatine, Iowa.*

*All of the requirements given were for a continuous flow process. Requirements used for the calculations of the boiler selection were based on the "worst case conditions" of the three processes considered.*

*Cooking (ICFAR) 45.8 lb. steam/bu corn.*
*Distillation (ACR) 180 lb. steam/bu corn.*

*Drying dark distiller's grains—optional (ICFAR) 155 lb. steam/bu corn.*

### Assumptions

*2.6 gallons of ethanol produced per bushel of corn.*

*144 gallons of ethanol produced per batch.*

*Two batches are run a week.*

*Ethanol is produced during the eighteen weeks of the winter season.*

*Cooker operates three hours per batch.*

*Distillation column operates eight hours per batch.*

*Boiler efficiencies:*
*0.6 for regular boilers.*
*0.8 is recommended by ACR.*

*1 BHP = 34.5 lb. steam per hour at 1.0 boiler efficiency.*

*1 BHP = 33,480 BTU per hour at 1.0 boiler efficiency.*

### Calculations

*Boiler selection is based on peak requirements since the cooker is operated for three hours while the distillation column is also operating. (Drying the distiller's grains isn't considered in the boiler selection.)*

## Calculations

*Steam requirements of the cooker:*

$$\frac{\text{45.8 pounds steam}}{\text{1 bushel corn}} \times \frac{\text{114 gallons}}{\text{1 batch}} \times \frac{\text{1 bushel corn}}{\text{2.6 gallons}} \times \frac{\text{1 batch}}{\text{8 hours}} = \frac{\text{251 pounds steam}}{\text{1 hour}}$$

Steam requirements of the distillation column:

$$\frac{\text{180 pounds steam}}{\text{1 bushel corn}} \times \frac{\text{114 gallons}}{\text{1 batch}} \times \frac{\text{1 bushel corn}}{\text{2.6 gallons}} \times \frac{\text{1 batch}}{\text{8 hours}} = \frac{\text{987 pounds steam}}{\text{1 hour}}$$

Peak steam requirements for a 0.6 efficient boiler:

$$\frac{\text{1,238 pounds steam}}{\text{1 hour}} \times \frac{1}{0.6} = \frac{\text{2,063 pounds steam}}{\text{1 hour}}$$

Boiler rating in boiler horsepower (BHP) that will be required:

$$\frac{\text{2,063 pounds steam}}{\text{1 hour}} \times \frac{\text{1 BHP-hour}}{\text{34.5 pounds steam}} = \text{60 BHP}$$

Peak BTU requirements:

$$\frac{\text{60 BHP} \times \text{33,480 BTU}}{\text{1 BHP-hour}} = \frac{\text{2,008,800 BTU}}{\text{1 hour}}$$

If you examine Fig. 6-13 carefully, you'll see that this farm distillery setup is just a somewhat more complex version of the other stills we've described. Yes, it's bigger, costs more to build, and needs more thinking out before you begin . . . but it'll turn out a great amount of high-proof ethanol, too. Most important, the chemistry of making ethanol doesn't change at all . . . and neither does the physics. If you can design your still to incorporate energy-efficient innovations (like using a beer preheater which is itself heated by the vaporized alcohol), you can save yourself a lot of effort . . . and money. And if you can figure out a way to lay out your still so that the liquids are *draining* from one station to the next, instead of needing to be pumped (to a large extent this is a question of available space), then you'll save even *more*. Figure your needs and resources, and then use your ingenuity . . . and you'll come up with the exact system that's right for you.

## SOLAR STILLS

In certain areas of the country, solar energy is a viable source of heat for distillation. In other areas it may not be worth the time to even think about it. The advantages seem obvious: Solar energy is free, unending, and nonpolluting. However, the disadvantages can sometimes outweigh the advantages unless one is in an area where the sun can be counted on to shine. For example, the mash has to be run when it has finished the fermentation process, or it will turn to acetic acid within a few days. There are certain chemicals that can be added to the mash to hold it for a while, but doing so only adds to the cost of fuel production.

To make a large amount of alcohol fuel, one would need to construct a considerable number of collectors. Also, solar panels seem to be very slow in production and low in product proof.

The research staffers of THE MOTHER EARTH NEWS® built a collector (see issue 56, page 114), according to Lance Crombie's specifications. The results were not very encouraging, though we *were* able to register increases in proof strength. However, for those who wish to pursue the quest, here are a few guidelines that the staff discovered which will make the solar still function better.

A solar still works best if the mash is preheated. A serpentine pattern copper pipe solar panel used

*COUNTERCLOCKWISE FROM UPPER LEFT: One of the first solar stills that we tested produced only 40-proof alcohol. . . . Neither the flat plate . . . nor the domed model proved better. . . . This conveyor belt unit did produce 90-proof "juice".*

*A promising approach to sun-powered distillation involves using a concentrating solar collector to generate steam that warms a heat transfer medium. Temperatures approaching 1600°F have been produced by this experimental collector.*

to preheat the mash before it reaches the still can raise the mash temperature to anywhere from 120 to 180°F (depending on the square footage of the panel). Then, as the hot mash enters the still, it will easily flash to a vapor. Remember that water is vaporized along with the alcohol, so this "steam" is not more than about 90 to 100 proof. Also, considerable alcohol is left in the discharge, so the "remains" will have to be recirculated to remove all the alcohol from the mash. The second running will have an even lower proof than that of the first batch.

Alcohol vapors rise, so—to prevent adding more water to the distillate—it's best to provide a means of removing the alcohol at the top of the still, not at the bottom. If the condensed alcohol vapors are allowed to run down the glass, they will gather the water droplets that have condensed on the lower portion of the glass and thus reduce the proof.

Perhaps one of the better methods of using solar energy is as a heat source in conjunction with a regular still . . . to preheat the mash (as described above) and thus reduce the amount of fuel required to bring it to a boil. Also, a large mirrored solar furnace (see THE MOTHER EARTH NEWS® NO. 55, page 93 . . . No. 56, page 142 . . . and No. 57, page 66) could make enough steam to operate a small distillery.

Much research will have to be done to make a solar still produce large enough quantities of alcohol fuel to warrant the cost of equipment. People with small fuel requirements may find it an acceptable method of production, but they might still wish to have a backup system just in case "ol' Sol" doesn't shine just when the mash is ripe.

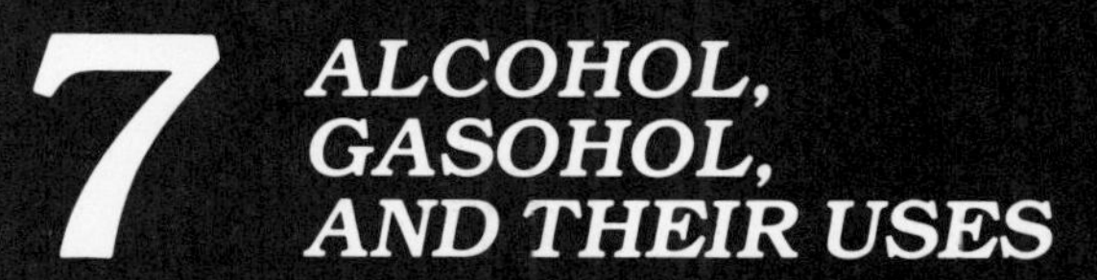


Now that you know how to make ethanol, you ought to learn what can be done with the fuel. Basically, you have two choices: Use it yourself . . . or sell it. By selling it, we mean trucking your batch down to a commercial distillery for conversion to pure-grade, 200-proof juice . . . or, perhaps, marketing it to (or bartering it with) your neighbors.

**CHART 7-1**

**PROPERTIES OF GASOLINE, ETHANOL, AND METHANOL**

| | GASOLINE* | ETHANOL | METHANOL |
|---|---|---|---|
| **Chemical Formula** | Complex | $C_2H_5OH$ ($CH_3CH_2OH$) | $CH_3OH$ |
| **Molecular Weight** | Complex | 46.07 | 32.04 |
| **Carbon to Hydrogen Weight Ratio** | 5.25:1 | 4.00:1 | 3.00:1 |
| **Carbon, % by weight** | 84.00 | 52.14 | 37.48 |
| **Hydrogen, % by weight** | 16.00 | 13.12 | 12.58 |
| **Oxygen, % by weight** | 0 | 34.74 | 49.94 |
| **Boiling Point, °F at 1 atmosphere** | 90 to 410 | 173.3 | 148.1 |
| **Freezing Point, °F at 1 atmosphere** | −65 to −161 | −174.6 | −144.2 |
| **Vapor Pressure, PSI at 100°F and 1 atmosphere** | 9.0 to 12.5 | 2.5 | 4.6 |
| **Latent Heat of Vaporization, BTU/lb. at boiling point and 1 atmosphere** | 141.00 | 378.13 | 481.00 |
| **Latent Heat of Vaporization, BTU/lb. at 77°F and 1 atmosphere** | 147.30 | 395.00 | 503.30 |
| **Heat of Combustion (Heating Value) at 77°F** | | | |
| **Lower Heating Value, BTU/lb.** | 19,000 | 11,550 | 8,600 |
| **Higher Heating Value, BTU/lb.** | 20,250 | 12,780 | 9,776 |
| **Lower Heating Value, BTU/gal.** | 110,960 | 76,230 | 56,760 |
| **Higher Heating Value, BTU/gal.** | 118,260 | 84,348 | 64,521 |
| **Weight per Gallon, lb.** | 5.84 | 6.60 | 6.64 |
| **Specific Gravity at 60°F** | 0.745 | 0.794 | 0.796 |
| **Flash Point** | −45°F | 56°F | 53°F |
| **Ignition Temperature** | 720°F | 685°F | 878°F |
| **Optimum Air/Fuel Ratio** | 15:1 | 9:1 | 6.5:1 |
| **Tolerable Air/Fuel Ratios, rich to lean** | 13.2:1 to 71.4:1 | 5.3:1 to 23.3:1 | 2.8:1 to 13.6:1 |
| **Octane Number, research** | 90 | 106 | 106 |
| **Maximum Desirable Compression Ratio, spark-ignition internal combustion engine** | 8.5:1 | 15:1 | 15:1 |
| **Energy of Optimum Air/Fuel Mixture, BTU/cu. ft.** | 94.8 | 94.7 | 94.5 |

***Since gasoline is a blend of octane and many different hydrocarbons, the figures given here are representative of an average high-test automotive gasoline sample.**

*"Old Green"—the first one of THE MOTHER EARTH NEWS® alcohol-powered vehicles—received a few simple carburetor modifications which enabled it to run on the renewable fuel.*

Or perhaps you'd prefer to use the juice you've brewed up to run your own vehicle(s), heat your own home, or whatever. That's for you to decide. What we're going to do in this chapter is show you some practical ways to use your home-brewed liquid energy . . . to make yourself less dependent upon the fossil-fuel market.

## ALCOHOL AS AN ENGINE FUEL

Before you begin to convert your gasoline-burning automobile or truck engine to use alcohol, it's important that you understand the properties of—and the differences between—the two fuels.

Gasoline is a complex mixture of hydrocarbons . . . substances comprising just hydrogen and carbon atoms. These hydrocarbons can appear in all forms (as a gas, a liquid, or a solid), but here we're concerned with the fuel in its liquid state.

To derive the various hydrocarbon fuels, the petroleum industry refines crude oil (made many millions of years ago as a result of geological and biological cycles) and draws off the desired product at a certain temperature and pressure. Hence, there are the lighter, gaseous fuels such as butane, propane, and ethane . . . the liquids like octane, pentane, and hexane . . . the heavier, oily liquids such as kerosene and fuel oil . . . and so on all the way down through waxes and solids.

Gasoline as we know it is a combination of octane, benzene, toluene, various other aromatics, tetraethyl lead, detergents . . . and compounds of sulfur, phosphorus, and boron. Because of the complexity of this mixture of ingredients—and because the refineries vary the blend to suit climate, seasonal changes, and altitude—it's difficult to choose a "representative" sample of gasoline for comparison purposes. Nonetheless, the figures that are given in Chart 7-1 (page 69) are fairly typical of average high-test automotive gasoline.

Alcohol, on the other hand, has to be manufactured . . . in our case through the fermentation and distillation processes. Because of the steps involved in its manufacture, alcohol has always been more expensive than gasoline to produce. But now, with dwindling crude oil supplies, the price of gasoline is skyrocketing. And soon (as finite petroleum supplies continue to dwindle) gasoline itself will probably have to be synthetically manufactured . . . and at a far greater expense than that required to produce ethanol fuel.

Alcohol compounds are also hydrocarbons . . . but in alcohol, one of the hydrogen atoms has been supplanted by a hydroxyl radical (hence the OH symbol), which is an oxygen atom bonded to a hydrogen atom (see Chapter 3, page 22 for the chemical formula of alcohol). Alcohols, too, take many forms and have various levels of complexity, but here we're concerned mainly with ethanol (grain-derived alcohol) and—just in passing—with methanol (wood- or cellulose-derived alcohol).

These two fluids are now the only common alcohol fuels . . . and of the two, ethanol is more economically feasible on a small scale. (The raw material used to make methanol—wood chips, garbage, or cellulose matter—is relatively inexpensive, but the manufacturing process necessary to produce methyl alcohol is, at present, economical only on an industrial level.)

On the surface, the difference between alcohol and gasoline might appear relatively minor: Alcohol contains oxygen, while gasoline doesn't. In reality, however, the dissimilarities are far more complex than that. And once the fuels are under compression—as is the case in an engine's combustion chamber—the comparison becomes even more complicated . . . but we'll get into more detail on these points later.

Regardless of the inherent differences between gasoline and alcohol, though, the fact is that alcohols make ideal motor fuels. Back in the mid-1800's —when coal gas was the most common motor fuel and gasoline was not even used—there was much experimentation with alcohol in engines . . . and the Model A Ford, produced from 1928 to 1931, was designed to burn a variety of fuels, one of which was alcohol. In addition, Studebaker trucks built for export in the 1930's (and various domestic tractors sold both in the U.S. and abroad) were offered with either gasoline or alcohol fuel systems. (Indeed, at the start of the "motorized era", alcohol was used just as commonly as—if not more so than—fossil fuels. But as time went on, segments of the auto and petroleum industries—which were organized and thus more powerful than small farm-based alcohol producers—lobbied successfully for the wholesale use of "superior" gasoline fuels. Strangely enough, in areas where petroleum had to be exclusively imported, or during times of war when gasoline supplies were rationed, alcohol suddenly became an excellent motor fuel again . . .

PHOTO COURTESY OF INDIANAPOLIS MOTOR SPEEDWAY

*Indianapolis-Class race cars have long run on alcohol. (Methanol, which powers the speedsters, is a cellulose-derived alcohol, and not as easily produced on a small scale as is ethanol.)*

*MOTHER's "Dual Fuel" (it runs on either alcohol or gasoline) van is the product of a far more extensive "conversion" than was "Old Green" (see page 70). The additional modifications allow the vehicle to extract more efficiency from the "new" fuel.*

and was often even touted as such by the petroleum distributors who were selling it!)

Be that as it may, alcohol has characteristics that make it a natural engine fuel: [1] It has a high "octane" rating, which prevents engine pre-ignition (knock) under load, [2] it burns clean . . . so clean, in fact, that not only are noxious emissions drastically reduced, but the internal parts of the engine are purged of carbon and gum deposits . . . which, of course, do not build up as long as alcohol is used as fuel, and [3] an alcohol-burning engine tends to run cooler than does its gasoline-powered counterpart, thus extending engine life and reducing the chance of overheating.

At this point, we can detail exactly how these and other characteristics of alcohol will affect engine performance.

## "Octane" Rating

Actually, when referring to alcohol fuels, the word "octane" does not technically apply at all, since octane (in its pure form) is merely the hydrocarbon in gasoline which is assigned the numerical value of 100 for fuel-rating purposes. The octane number given automotive fuels is really an indication of the ability of the fuel to resist premature ignition within the combustion chamber. (Premature ignition, or engine knock, comes about when the fuel/air mixture ignites spontaneously toward the end of the compression stroke, as a result of intense heat and pressure within the combustion chamber. Since the spark plug is supposed to ignite the mixture at a slightly later point in the engine cycle, pre-ignition is undesirable, and can actually damage or even ruin an engine.)

Because a high compression ratio in an engine results in more power per stroke, greater efficiency, and better economy, it's easy to see why a fuel that resists pre-ignition even under high compression conditions is especially desirable . . . and alcohol is, on the average, about 16 points higher on the research octane scale than is even premium gasoline.

## Heat Value

The heating value of a fuel is a measure of how much energy we can get from it on a per-unit basis, be it pounds or gallons. When comparing alcohol to gasoline—using this "measuring stick"—it's obvious that ethanol contains only about 63% of the energy that gasoline does . . . mainly because of the presence of oxygen in alcohol's chemical makeup. But since alcohol undergoes different changes as it's vaporized and compressed in an engine, the outright heating value of the ethanol isn't too important when it's used as a motor fuel.

The fact that there's oxygen in the alcohol's structure means that this fuel will naturally be "leaner" than gasoline fuel. This is one reason why we must enrich the air/fuel mixture (add more fuel) by increasing the size of the jets in the carburetor when burning alcohol . . . a conversion which we'll discuss in detail later in this chapter.

## Volatility

The volatility of a fuel refers to its ability to be vaporized. This is an important factor, because if vaporization doesn't occur readily, the fuel can't be evenly mixed with air and is of little value in an engine. Some substances that are highly volatile can't easily be used as motor fuels . . . and others, which have excellent heating value, aren't volatile enough to be used in an engine (examples of the latter group include tars and waxes).

Another point to keep in mind is that a very volatile fuel is potentially dangerous, because of the chance of explosion from heat or sparks. This is one reason why alcohol, with a higher flash point than gasoline, is a much safer automotive fuel . . . especially considering that the average car's fuel storage tank is—as we've come to learn—really quite vulnerable.

## Latent Heat of Vaporization

Latent heat of vaporization is the phenomenon that results in the ability of an alcohol-powered engine to run cooler than its gasoline-fueled counterpart. When a substance is about to change form, such as from a liquid to a vapor, it must absorb a certain amount of additional heat from its surroundings in order for the change to take place. Since alcohol must absorb roughly 2-1/2 times the amount of heat that gasoline does, *and* since that heat must be extracted from the engine block, the engine of an alcohol-fueled vehicle should operate at a much lower temperature . . . in theory, that is.

What actually happens is that the alcohol/air mixture doesn't have time to absorb all the heat it could during its short trip through the engine manifold. So instead of running *very* much cooler on alcohol than it does on gasoline (which, by the way,

*A pair of Carter carbs are used on the dual-fuel van. One of them is plumbed for ethanol, one for gasoline . . . and the operator of the vehicle can switch from carb to carb easily.*

*Under emission-test conditions, a New York taxicab—sporting inexpensive carburetor modifications and burning homemade alcohol—scored a remarkably low carbon-monoxide "rating".*

would not be desirable . . . since an engine must retain a certain amount of heat to run efficiently), the engine operates at temperatures only *slightly* cooler—about 20–40°F lower, depending on the specific engine—when using alcohol fuel.

## Exhaust Emissions

When gasoline is burned in an engine, it produces carbon monoxide and other poisonous fumes . . . both because of the fact that the fuel fails to combust completely and because it's subjected to extreme temperatures and pressures. In addition (as we mentioned before) gasoline is a complex mixture of many compounds . . . and some of those substances are lead, sulfur, and other noxious materials. These, too, add to the contaminative effects of the engine's exhaust fumes.

Alcohol, on the other hand, burns much cleaner. Even though it, too, never combusts completely, the volume of noxious fumes produced is drastically reduced in an alcohol-burning engine . . . because alcohol contains oxygen in its structure (which results in a more thorough combustion) and *doesn't* contain all the other pollutants necessary as additives in gasoline.

For comparison purposes, MOTHER's research crew ran tests on a 1978 Chevrolet taxicab . . . which, operating in New York City, was subject to some of the most stringent pollution controls in the nation. (In order for cabs to be licensed, they must undergo—and pass—four scheduled EPA tests a year for carbon monoxide emissions and several more for hydrocarbon.) Naturally, the taxi that MOTHER's crew tested was much less polluting than is the average American automobile, but even in perfect tune it just "squeaked by" the tests using gasoline . . . registering nearly a 1-1/2% carbon monoxide (CO) and a 200-parts-per-million hydrocarbon (HC) exhaust content (both just under the legal limit).

With alcohol fuel, however, the test results improved enormously. Even with all pollution controls removed from the engine (except for the PCV valve), the cab registered a mere 0.08% CO and only 25 PPM of HC . . . the equivalent of 95% less CO and 87.5% less HC!

## Alcohol/Water Mix

As we all know (some of us from experience), water and gasoline don't mix. The gasoline tends to float to the top of the mixture, leaving the water to settle below it. In a car's fuel tank, such separation can be disastrous, particularly during the winter season.

Alcohol, however, mixes quite well with water: The water molecules distribute evenly within the mixture. As a result, not only is the winter freezing problem solved, but there is an added economy bonus because *pure* alcohol is not necessary for fuel purposes. This is very important to the small-scale alcohol fuel producer, since nonindustrial stills are generally not capable of producing more than 192-proof (96% pure) alcohol.

As far as its use in an engine is concerned, MOTHER's team of researchers has had excellent results with various strengths of alcohol, from 160 to 200 proof. (More water added beyond the 20% limit causes the engine to hesitate and idle roughly . . . hence that practice is not recommended.) As an extra benefit, the water in the fuel helps cleanse and "lubricate" the internal parts of the engine, including the valve seats, piston head surfaces, and combustion chamber.

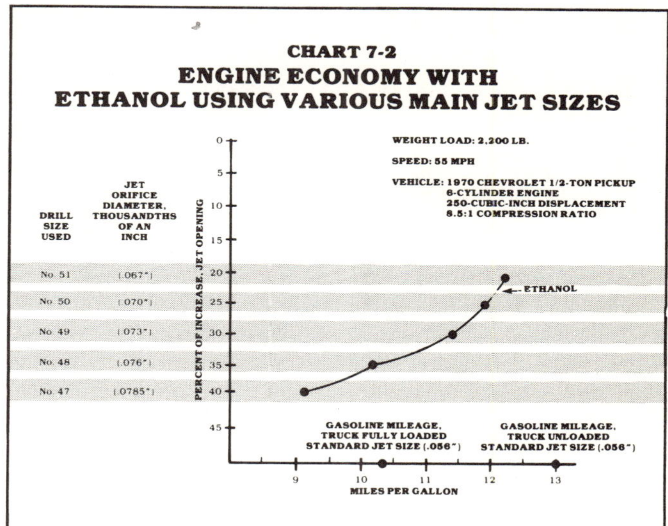


## Engine Economy

The fuel economy of a particular engine is directly related to the ratio of the air/fuel mixture . . . and that, of course, is dependent upon how large the main jet in the carburetor is. Alcohol requires a richer air/fuel mix than does gasoline (9:1 as opposed to 15:1), but that difference is not reflected proportionately with respect to economy . . . partially due to the fact that alcohol has a higher "octane" rating and can be utilized more efficiently.

By experimenting with orifice sizes (diameters) of the carburetor's main jet, it is possible to reach a happy medium between power and economy when using alcohol in a standard auto engine. Chart 7-2 shows that a 40% increase in diameter over the standard jet size in MOTHER's truck resulted in a loss of only 12% of the total fuel economy . . . as compared to burning gasoline and using the standard jet (the truck was fully loaded in both cases).

With a step-by-step decrease in jet size, down to 19% larger than the original, only about 5% was lost when compared with the *unloaded* truck using gasoline and its normal jetting. When that test was made comparing both fuels in the fully loaded vehicle, however, the alcohol-powered version actually showed a relative *increase* in mileage . . . by a

*A "FloScan" meter—which will give accurate mile-per-gallon or gallon-per-hour figures while a vehicle is being operated—was used in much of MOTHER's alcohol experimentation.*

whopping 16%! This is because an alcohol-powered vehicle—partly as a result of the fuel's high "octane" rating—will *maintain* its economy even under extreme loads, unlike most gasoline-powered cars. The figures given above were recorded while MOTHER's 1/2-ton pickup was pulling over 2,200 pounds . . . and only improved slightly when the load was removed. In comparison, when the weight was removed from the truck in its gasoline mode, mileage improved substantially. (See Chart 7-2.)

It should also be noted that an increase in engine compression ratio will improve alcohol mileage considerably . . . perhaps to the point where, loaded or unloaded, any vehicle could equal or better its gasoline fuel mileage.

### Engine Performance

An engine powered by alcohol—if converted correctly—will have performance equivalent to, if not greater than, the same powerplant burning gasoline. This is because of the fact that alcohol has a higher "octane" rating (hence the timing can be advanced slightly), and can stand much greater compression ratios.

Even without changing the compression ratio, an alcohol-powered engine with fairly low compression (MOTHER's pickup has a ratio of 8.5:1) still holds its own against its gasoline-burning counterpart. And, if the timing is advanced safely short of the "knock" limit, the torque range is broadened considerably, allowing the vehicle to pull under load exceptionally well . . . in fact, it does much better under such conditions than does the gasoline version!

## HOW TO ADAPT YOUR AUTOMOBILE ENGINE FOR ETHYL ALCOHOL USE

Now that we've explained the fundamental differences between alcohol and gasoline fuels, we can get on with the actual conversion of a conventional gasoline-burning engine to alcohol use. We'll cover the three major changes (main jet, idle jet, and timing), and we'll also go on to describe some other areas that may be of interest to those who want to go further to increase the efficiency of their alcohol-burning engines.

### Main Jet Changes

The first thing you'll have to alter is the main metering jet in your carburetor. In most carburetors, this is a threaded brass plug with a hole of a specific size drilled through its center. This hole is called the *main jet orifice*, and its diameter dictates how rich or lean the air/fuel mixture will be when the car is traveling at cruising speeds. Now, the smaller the hole is, the less fuel will blend with the air and the leaner the mixture will be . . . but as the orifice is enlarged, more fuel is admitted, and the mixture is enriched.

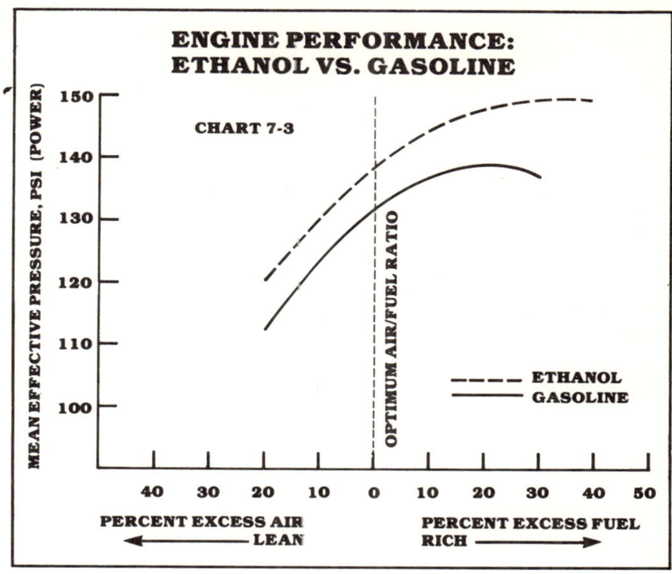


Since alcohol requires a richer air/fuel ratio, it's necessary to bore out the main jet orifice when using ethanol fuel. The standard jet size in MOTHER's alcohol-powered truck was .056" . . . in other words, this was the diameter of the jet orifice. In order to operate the engine successfully on alcohol fuel, it's necessary to enlarge this opening by anywhere from 20 to 40%

Start your conversion by gathering all the tools and hardware you'll need to complete the job. A screwdriver, an assortment of end wrenches, Vise-Grip pliers, a putty knife, a pair of needle-nose pliers, and a power drill—with bits ranging in size from a No. 51 (.067") to a No. 46 (.081")—are usually all you'll require. To make your job easier, though, you might want to refer to a Motor, Chilton, or Glenn auto repair manual for exploded illustrations to guide you through the necessary carburetor disassembly and reassembly. (A second alternative would be to purchase a carburetor rebuilding kit for your make and model car . . . which will not only supply you with a working diagram, but provide gaskets, seals, and other parts that may get damaged during the strip-down process.)

You may also need to purchase several main jets from your auto dealer (if the carburetor you're converting has a removable main jet), since you'll probably want to experiment with different air/fuel ratios.

In order to take the carburetor apart, you'll first have to remove its air-filter housing and all its hoses, tubes, and paraphernalia from the engine. Then disconnect the throttle linkage from the engine and any choke linkage rods that aren't self-contained on the carburetor body. (If you've got a manual choke, remove its cable and tie it out of the way.)

You'll also have to unscrew the fuel line from the carburetor inlet fitting and remove any other hoses that fasten to the unit, including vacuum and other air-control lines.

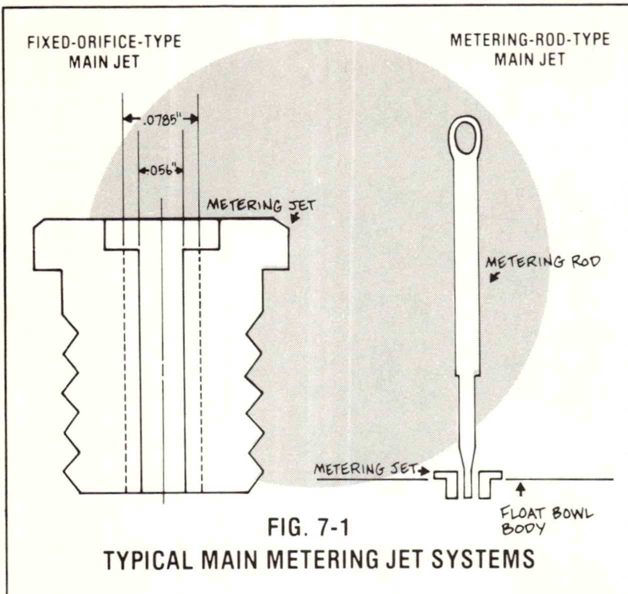


FIG. 7-1
TYPICAL MAIN METERING JET SYSTEMS

When the carburetor is free from all external attachments, remove it from the manifold by loosening the hold-down bolts at its base, and turn the unit upside down to drain out any gasoline that may be in the float bowl. Remove the carb's air horn (you may have to unfasten the choke step-down linkage rod) and locate the main jet. (In some carburetors the jet is installed in a main well support, while in others the brass fixture is mounted right in the float bowl body.)

Once you've removed the main jet, you can prepare to enlarge it. First measure the diameter of its opening by slipping a drill bit of known size into the hole (this bit should fit snugly, of course). In some cases, the size of the jet is stamped in thousandths of an inch right on its face, so you don't have to go to this trouble. When you know what the standard jet size is, you can calculate the additional enlargement necessary to operate the engine on alcohol.

For example, MOTHER's truck originally had a .056" main metering jet. In order to increase that opening's diameter by 40%, we first had to multiply .056" by .40 (which yielded .022"), then we had to add that additional .022" to the original .056" . . . this figured out to a total diameter of .078". The nearest size drill bit to this is a No. 47, which is .0785" in diameter . . . this was the bit necessary to bring the jet to 40% over its original size.

After you've determined the size of the drill bit needed to enlarge your vehicle's main jet orifice, hold the jet with your Vise Grips and *carefully* bore out its central hole (if possible, use the jet-holding body of the carburetor itself as a mount while you drill). Be sure to do your drilling as nearly straight as possible, and clean any brass residue out of the carburetor and its components after the operation is over.

There are some carburetors that not only use a fixed-size jet but also utilize what is known as a "metering rod". (For diagrams of both types of jets, see Fig. 7-1.) This is usually a thin tapered or stepped brass rod that's suspended within a brass jet orifice, which may or may not be removable. The fuel is, in this case, drawn through the space between the rod and its brass "housing". Depending on how far the throttle is opened, the metering rod is lifted out of the hole . . . and—since the rod is thick at its "base" (near the top), and progressively thinner at its tip (toward the bottom)—the farther it's drawn out of the hole, the more fuel is allowed to flow between the central rod and the opening.

The conversion on this type of metering system is basically the same as the fixed-jet conversion. To enlarge the orifice, you can either remove the metering rod and very carefully drill out the brass jet (take it out of the carburetor if it's removable), or take the tapered brass rod to a machine shop and have it turned down slightly (the same effect can be accomplished less accurately by sanding the rod down with emery cloth). If you choose to drill the jet to a larger dimension, the diameter should be increased anywhere from 10 to 32%.

With the fixed-jet type of carburetor, the diameter of the jet orifice can vary from about 20% over standard to 40% larger—or even more—depending on the engine's size, its compression ratio, and the vehicle's weight. Probably the best way to determine what is right for your needs is to experiment, since many instruments used to measure the proper air/gasoline ratio don't register correctly when the engine is burning alcohol.

By planning on a diameter enlargement of anywhere from 35 to 40% at first, you'll be perfectly safe, since the engine will tolerate this size easily. If you go too much larger than this, you'll probably just be wasting fuel. But by going too small, you may find that you'll lose power . . . or even worse, that you'll burn your vehicle's valves because of an overly lean mixture.

On the other hand, it *is* true that a *slightly* lean mixture—if carefully monitored—will result in improved economy with hardly a noticeable loss in performance. With MOTHER's vehicle, the absolute minimum was a 19% enlargement in jet size . . . although the truck does run a bit better with a 25% larger-than-standard main jet. You may find, as we did, that your vehicle performs well with a smaller jet opening than the suggested 35–40% increase . . . but to be on the safe side, periodically examine your spark plugs, especially after an extended drive. If they are white in color, or otherwise appear to be subject to excessive heat (look for hairline cracks on the center electrode's insulative jacket), this is an indication that your engine is burning too lean . . . and the jet must be enlarged.

### Idle Orifice Changes

Most carburetors will require additional idle circuit enlargement in order for the engine to run

*TOP: In MOTHER's gasoline-to-alcohol carburetor modifications, the float bowl assembly is removed and the brass main jet is unscrewed. ABOVE: It's best to use part of the carburetor itself as a "mount", while enlarging the main jet by about 40%.*

at the slowest, or idle, speeds. This is because the circuit fed by the main jet operates fully only when the throttle plate within the throat of the carburetor is opened past the idle position. When the plate is in the idle position, the air/fuel mixture is allowed to enter the manifold only through the idle orifice itself . . . which, if it isn't large enough, will not provide the needed amount of air/fuel blend to keep the engine running.

On some engines, it may be necessary only to loosen the idle mixture screw at the base of the carburetor in order to provide the correct amount of fuel, since this threaded shaft has a tapered tip which allows more mixture to pass as the tip is backed off. On other engines, it's possible that the seat itself, into which the tapered screw extends, must be enlarged in order to accomplish the same results.

In most cases, if the seat has to be bored out, it can be enlarged by 50%, using the same method of measurement as was detailed in the main jet section. This will allow a full range of adjustment with the idle mixture screw, even if you should want to go back to gasoline fuel. (When drilling, be careful not to damage the threads in the carburetor body.)

As a precaution against the idle screw's vibrating loose from its threaded opening, you can shim the idle mixture screw spring with a couple of small lock washers . . . this will prevent the screw from turning even if it's drawn out farther from the seat than it normally would be.

## Power Valve Changes

Most modern auto carburetors have a power valve that allows extra fuel to blend with the air/fuel mixture when the accelerator is depressed, in order to enrich the mixture under load conditions. This vacuum-controlled valve is spring loaded, and shuts off when it isn't needed . . . in order to conserve fuel.

The power valve used in the carburetor illustrated is somewhat difficult to alter and, besides, is sufficient for alcohol use in its normal configuration if it's working properly. However, there are other carburetors—specifically the Holley and Ford (Autolite or Motorcraft) brands—that have easily replaceable power valves which are available from auto parts stores in various sizes. If you use a power valve with a 25% or so greater flow capacity than the one that originally came with the carburetor, your air/alcohol mixture will be sufficiently enriched when your engine needs more power.

### Accelerator Pump Changes

In addition to a power valve, almost all automotive carburetors utilize an accelerator pump. This is a mechanically activated plunger or diaphragm that injects a stream of raw fuel directly down the throat of the carburetor when the accelerator is suddenly depressed. The fuel is injected through a small orifice located in the throat wall at some point above the carburetor venturi (the point at which the throat narrows).

The reason the accelerator pump is incorporated into modern carburetors is that—as the accelerator is pressed and more of the air/fuel mixture is drawn into the cylinders—some of the liquid particles in the blend tend to stick to the walls of the intake manifold, effectively leaning out the mixture by the time it reaches the combustion chambers. The extra squirt of fuel that's added by the accelerator pump makes up for this initial lean condition.

In order to adapt your accelerator pump to use alcohol effectively, you'll probably have to enlarge the size of the injection orifice slightly. (Anywhere from 10 to 25% is fine . . . if you go larger than that, you'll risk the possibility of altering the pump pressure enough either to turn the fuel stream into a dribble or to empty the pump reservoir before the pump has made a full stroke.)

As an alternative to enlarging the hole, you may be able to simply adjust the stroke length of the pump arm in order to feed more fuel. Most carburetors installed on Ford products already have a provision for seasonal adjustment, so—in such vehicles—the "change" is just a matter of putting the pump on its richest setting. Other carburetors, too, have threaded rods that can be adjusted to achieve the same results.

### Choke Alteration

Although it's not absolutely necessary to adapt your car's choke system when the vehicle burns alcohol fuel, it has been our experience that a manually operated choke is more desirable on an alcohol-powered car. If your vehicle's engine is already so equipped, fine. If not, you can purchase—for about $7.00 from any auto parts store—a manual choke conversion kit that will allow virtually any automatic choke to be adapted for manual control.

### Ignition Timing

In order to take advantage of the great antiknock qualities that alcohol fuel provides, you'll have to advance the engine's ignition timing by turning the distributor housing *opposite* to the direction in which the rotor spins (the housing is held in place by a bolted clamp).

Normally, an engine using gasoline has its timing set so the spark occurs at anywhere from 8° BTDC (Before Top Dead Center) to TDC (Top Dead Center). Since alcohol has a higher "octane" rating, you can advance the timing considerably more than this. In the case of MOTHER's truck, we adjusted it to operate at approximately 22° BTDC without any sign of pre-ignition, even under load. Of course, care should be taken when you adjust the timing on *your* vehicle, since a 22° advance might be excessive for your car. Remember, it's not safe to be *just* short of "pinging", since inaudible knocking can also damage the engine . . . the best procedure is to set the distributor timing at least two degrees retarded from the point of pre-ignition.

### Compression Ratio Changes

Increasing the compression ratio of the engine will be impractical for most people, because of the expense and work involved . . . however, this modification will do a great deal to improve engine performance and economy. Just as will a timing advance, a compression ratio hike will allow your engine to take advantage of the potential that alcohol has to offer as a fuel. Optimally, the ratio can be increased to 14:1 or 15:1 . . . but even a nominal increase—to perhaps 10.5:1, a figure that some manufacturers have offered in the past for premium gasoline use—will result in a vast improvement over the standard 8:1 or 8.5:1 that most manufacturers incorporate into their engines today.

If you intend to convert an automobile that already has a compression ratio of 10:1 or better, it probably won't pay to make any internal changes. However, if the engine you're considering needs an overhaul, it would be wise to modify it regardless of its compression ratio.

The most inexpensive way to increase your compression ratio is to install a set of high compression pistons. The forged units are designed to pack the air/fuel charge tightly into the combustion chamber for increased power, and have special relief notches built into their heads for valve clearance.

Be cautioned, however, that some engines may not tolerate a 15:1 compression ratio with standard connecting rods and bearings. These components, too, may have to be replaced with high-strength, competition-grade parts.

Another way of increasing the compression ratio slightly is by "milling" (planing) the surfaces of the cylinder head and/or block. With some engines, this may result in only a 1/2-point ratio increase . . . with others, slightly more. It would be best to check with your local engine rebuilder or automotive machine shop to determine exactly what—in terms of performance and mileage—you'll gain with your particular model engine *before* you go to the trouble of dismantling it.

A third—and perhaps the most versatile—way of effectively increasing the compression ratio is by installing a supercharger or turbocharger. These units, although ranging in price from $800 to over $1,200, provide a pressure boost in the combustion chamber proportional to the engine's load and/or RPM. Hence, compression would not be excessive during engine start-up as it would be with the other methods.

You should encounter no problem with a severe compression ratio increase, unless you decide to switch back to gasoline fuel . . . and in such a case, you could simply install a water injection system that would allow you to operate the car on regular fuel without fear of pre-ignition.

## Fuel Preheating

In extremely cold climates, it may be necessary to preheat your alcohol fuel before it enters the carburetor float bowl. This can be accomplished by splicing into the fuel feed line—near the point where it passes the upper radiator hose—and installing a fuel heater at this location.

You can fabricate a fuel heater in a matter of minutes by first locating a 5″ section of copper or other metal pipe with an outside diameter equal to that of the inner diameter of your upper radiator hose. Then find several feet of soft copper tubing that will slip snugly inside your fuel feed line. (If your fuel line is steel, you'll have to cut it and splice in two short sections of the appropriately sized neoprene hose.) Wrap the soft tubing several times around the middle of the large pipe section (the number of coils depends on how warm you want the fuel to become, but anything from three to

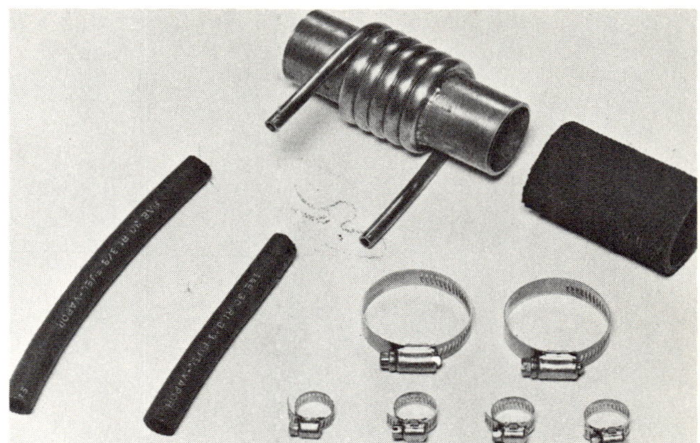

*A section of pipe, a few feet of soft copper tubing, a number of hose clamps, and a length of hose are all the components that are necessary to assemble an alcohol fuel preheater.*

*The preheater is placed between the existing radiator hose and another short section of tubing connected to the radiator neck (see the completed installation on the following page).*

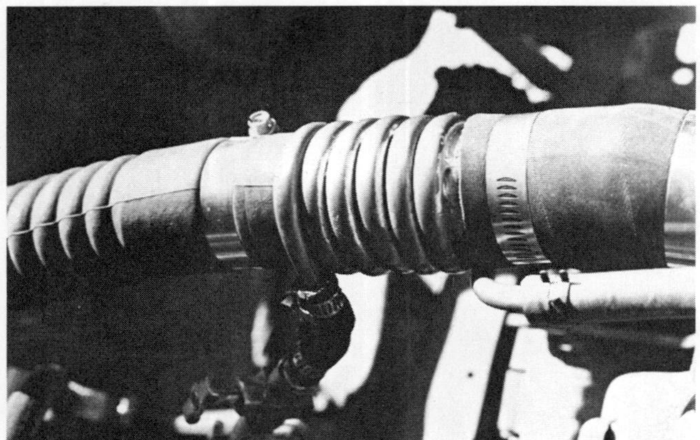

*In place, MOTHER's preheater will provide better fuel economy and smoother running for alcohol-powered vehicles in cold climates. (Do not use a preheater on a gasoline-powered vehicle.)*

eight wraps will suffice), and solder it in position.

To install the unit, just clamp it in place between the existing radiator hose and another short section of hose connected to the radiator neck, and attach the fuel line to the inlet and outlet of the copper coil. As the engine reaches operating temperature, the hot water flowing through the engine's cooling system will heat the coils and the fuel passing through them.

## Air Preheating

Most trucks and autos have air filter housings which are designed to allow heated air from around the exhaust manifold to channel through a duct and enter the carburetor when the engine first starts from a cold state. As the engine warms up, a flap within the air cleaner "snorkel" shuts off this supply of warm air and allows ambient air from the engine compartment to enter in its stead.

This flap is usually either thermostatically or vacuum controlled . . . but either way, you may find it helpful during the winter months to leave this valve closed to the cold outside air. This can be done either by disconnecting the bimetallic thermostat spring that controls the flap and installing a small spring of your own that will hold the valve in the required position, or—if the flap is vacuum activated—by connecting an existing permanent vacuum line to its control fitting. (As an alternative, you could remove the control line entirely, plug it up, and hold the flap closed with a spring.)

## Thermostat Change

In order to get maximum efficiency from your engine, you may need to change the thermostat within the engine block. Thermostats are available in various heat ranges from 140 to 200°F, and these temperatures indicate just how hot the engine coolant will be allowed to get before the thermostat opens to initiate the cooling process. (A thermostat is designed to hold the coolant within the cylinder head till it achieves the desired temperature . . . at which point the heated liquid is allowed to escape into the radiator to be cooled, and is replaced by a fresh supply of cool fluid. Depending on the engine's operating conditions, the thermostat may cycle open and shut regularly over the span of a few minutes.)

If the water in your vehicle isn't getting warm enough to provide hot air through the heating system, you should replace the thermostat with a higher-rated unit. By the same token, the intake manifold of your engine should be warm to the touch when burning alcohol. If it's cold—or iced over—the alcohol most likely isn't being given a chance to vaporize sufficiently and, therefore, is not being used efficiently. By using a hotter thermostat, you'll be able to warm up the entire engine, including the intake manifold.

## Cold Weather Starting

Since alcohol doesn't vaporize as easily as does gasoline, cold weather starting can be a problem . . . especially if the engine itself is cold. To alleviate this undesirable situation, MOTHER's research staff has designed a combination cold-start/dual-fuel system that'll work with any car.

All it requires is a five-gallon fuel storage tank with a fuel filler neck brazed into its top (we used an old propane bottle), an auxiliary electric fuel pump, some steel brake or fuel line, neoprene hose, an elbow, a length of copper pipe, a small metering jet, and several needle valves, tees, and hose barbs. (See Fig. 7-2.)

The five-gallon tank is mounted in some safe place on the truck or automobile and used to store gasoline. This cache of petroleum fuel serves a dual role: When it's needed for cold starting purposes, the electric pump is activated momentarily from inside the car and a fine stream of gasoline is injected down the throat of the carburetor. And, in the event that your alcohol supply is unexpectedly depleted on the highway, the gasoline stored in the small tank can be routed into the carburetor's normal circuits for emergency use.

## Initial Use of Alcohol Fuel

An engine altered as outlined in this chapter will run well on alcohol. Nonetheless, there are certain things to be aware of as you begin to make use of the new fuel. First, remember that the alcohol will act as a cleansing agent . . . and—as such—will not only clean out your tank, fuel lines, and filters, but will also purge your engine's internal parts of built-up carbon, gum, and varnish deposits.

In effect, what this means is that suddenly a lot of filth will be floating around in your fuel . . . and it may be enough to clog your fuel filter to the point of not allowing any fuel to pass. By the same token, loosened internal engine deposits can foul the

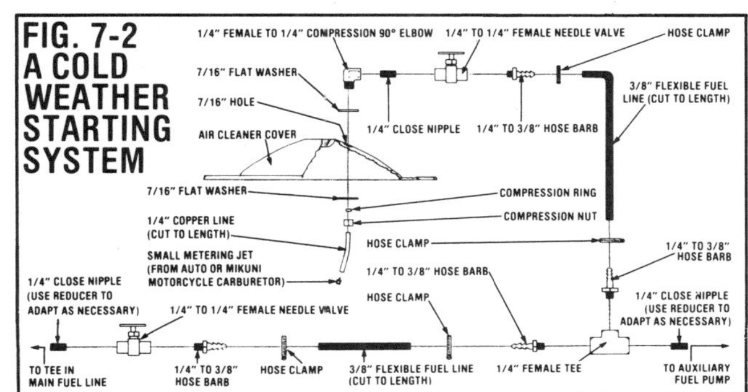


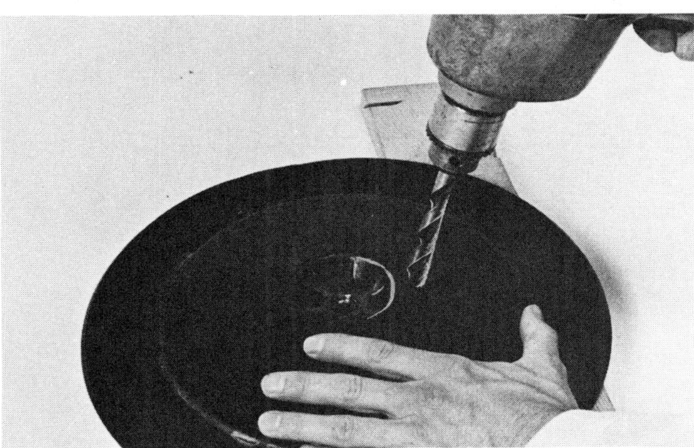

*FROM THE TOP DOWN: The cold weather starting system's fuel line enters the air cleaner through a 3/8" hole. . . . The finished injector assembly is attached to the air cleaner cap.*

**CHART 7-4**
**DURABILITY OF VARIOUS PLASTICS: ALCOHOLS VS. GASOLINE**

| | Ethanol | Methanol | Gasoline |
|---|---|---|---|
| Conventional Polyethylene | good | excellent | poor |
| High-Density Polyethylene | excellent | excellent | good |
| Teflon | excellent | excellent | excellent |
| Tefzel | excellent | excellent | excellent |
| Polypropylene | good | excellent | fair |
| Polymethylpentene | good | excellent | fair |
| Polycarbonate | good | fair | fair |
| Polyvinyl Chloride | good | fair | poor |

Excellent: Will tolerate years of exposure.
Good: No damage after 30 days of exposure . . . should tolerate several years of exposure.
Fair: Some signs of deterioration after one week of exposure.
Poor: Deteriorates readily.

NOTE: All tests were made with liquids at 122°F.

spark plugs badly . . . so if your vehicle begins to function poorly soon after your conversion, check these two areas first.

In addition to the fact that alcohol is a cleansing agent, it is also a solvent . . . and this means that certain types of plastics used in the fuel system of your vehicle *may* be attacked by ethanol. Actually, most deterioration of plastics associated with ethanol fuel is caused by the substances used to denature it—such as acetone or methyl ethyl ketone—rather than by the alcohol itself. If you manufacture your own alcohol and denature it with gasoline, deterioration problems will be reduced to a minimum.

Most vehicles manufactured prior to 1970 used stainless steel or brass components in their fuel systems . . . hence there is little chance of parts failure in such cars or trucks. In autos that use plastic components, however, there are several areas of potential deterioration: [1] Within the fuel tank, both the float and the strainer on the fuel intake tube may be plastic . . . replace them if necessary. [2] The fuel lines themselves—if they are the clear, flexible type—may also soften . . . you can install neoprene hose in their place. [3] The fuel pump diaphragm may also be subject to failure . . . either replace it with a piece of spring steel, or replace the entire pump with an electric gear-type model available from your auto parts store. (Jaguars and Alfa-Romeos also use all-metallic pumps if you're willing to pay the price to buy and adapt such a pump to your vehicle.) [4] Plastic in-line fuel filters should be replaced with metal ones. [5] Many modern carburetors use plastic float needles, seals, and floats . . . you can usually purchase the equivalent carburetors—but ten years older—from an auto wrecking yard for about $5.00. These should contain metal components, and can be salvaged for parts.

Of course, not all plastics are subject to corrosion, and neither are all types of rubber. Generally, butyl rubber (like the type used in inner tubes) should be avoided. Neoprene, however, holds up well even at higher temperatures, and will present a problem (because of swelling) only if it's used as a tip on carburetor float needles. Automotive plastics vary greatly in their composition . . . Chart 7-4 indicates the durability of a selection of alcohol-tolerant plastics.

One final thing to be aware of when burning alcohol in your vehicle is that the new fuel does not contain the additives which the engine has become used to over the years . . . specifically the leads which help to lubricate the valve seats. Of course, any car built in 1975 or later is already equipped with hardened valves and seats, so there should be no problem with them . . . but even vehicles of other years (with the possible exception of large-block 1972–1974 Ford products) can tolerate alcohol fuel safely.

One reason for this is that water in the alcohol acts as a "cushion" and lubricant for the valves . . . but if you are still wary of using alcohol fuel in its pure form, you can add up to 1% kerosene or diesel fuel to your alcohol supply. This will provide the lubrication of petroleum fuels with a minimum of pollution.

### Fuel Injection Systems

Since some vehicles are equipped with fuel injection rather than carburetors, we will briefly touch on the use of alcohol with that system. There are two important factors in a fuel injection setup: injection timing and control jet diameter. Fortunately—since many systems now use an electronically controlled timing sequence—injection timing is not critical in an alcohol-injected engine. Neither performance nor economy will be improved substantially by either advancing or retarding the injection timing process.

Control jet diameter, on the other hand, is an important adjustment factor. If you increase the size of the control jets (which are the equivalent of the metering jets in a carburetor), the engine will operate well on alcohol fuel. An increase of 15–20% is all that's necessary to accomplish the conversion. (Ignition timing should, of course, be advanced as explained previously.)

An interesting feature of the fuel injection system is that it doesn't require any gasoline during the cold weather starting process to fire the engine up. Since the fuel is injected under pressure, the alcohol fuel is sufficiently vaporized to ignite easily within the combustion chamber.

### Diesel Engines

Because of the fact that diesel engines do not use conventional spark ignition systems, it's difficult for pure alcohol to ignite within the combustion chambers. This, coupled with the fact that diesel

*This Lombardini diesel engine has been successfully run on "diesohol" mixtures of up to 15% ethanol. (The conversion of diesel engines to alcohol fuel alone is still under study.)*

injector pumps won't tolerate water, could be a problem . . . especially if the alcohol used is not nearly pure.

Fortunately, there are several other ways to utilize homemade ethanol in a diesel engine by introducing vaporized alcohol to the engine along with diesel fuel. Probably the simplest way is to mount an automobile carburetor right on the diesel's air intake manifold and supplement the diesel fuel with alcohol metered through that piece of equipment. Of course—just as in a conventional gasoline engine—as incoming air rushes down the air inlet tube, it will pick up alcohol vapor metered through the carburetor . . . which should have a controllable throttle to match tractor load.

Another way to use ethanol in a diesel engine is to install fuel injectors into the intake manifold to accomplish the same result. This system would require a separate pump that would have to be timed in order to inject alcohol at the proper moment.

A vaporizer—like those found on propane fuel systems—can also be used to add alcohol to the diesel fuel system. This, again, will provide the diesel intake manifold with ethanol vapors that help combustion.

Since a diesel engine has closer tolerances—and is more costly to repair—than a conventional gasoline engine, you should take extreme care when altering and running diesel equipment on other than pure diesel fuel. If you don't consider yourself competent to work on diesels, find someone who is . . . since the diesel fuel injector pump must be adjusted to provide less flow when alcohol fuel is used, plus the fact that a lean mixture condition—and even increased horsepower outputs—can damage a diesel engine in short order.

Turbocharged diesels can be equipped with what is known as an "aquahol" injection system, to be marketed by the M & W Gear Company of Gibson City, Illinois early in 1980. This setup injects a fine mist of alcohol and water in a 1:1 ratio directly into the engine's air intake, which results in a lowering of fuel consumption and a tolerable increase in horsepower.

## GASOHOL

Of course, most of the talk around the country about using alcohol fuel for cars has centered on using a gasoline/ethanol blend . . . *gasohol* as it's called. Usually that term means a 9:1 (sometimes

4:1) mix of gas and anhydrous alcohol (the 200-proof fuel that it takes a commercial distillery to produce). Gasohol requires the ultra-pure ethanol because the water in lower-proof alcohol won't mix with gasoline . . . and when it separates out and gets to the engine, the car stops dead. So using gasohol isn't of any advantage to the person who wants to make his or her alcohol fuel . . . at least not so far as "making what you use" goes. There *are* some definite general advantages to the mix, though.

First, a standard automobile or truck *gasoline* engine will run on gasohol with no modifications whatsoever . . . some say more efficiently. (The results of various tests run to see if gasohol is more fuel-efficient or less polluting than gasoline have been contradictory and therefore are inconclusive.) And, of course, for every ten gallons of *gasohol* burned in place of conventional fuel, one less gallon of *gasoline* is being used. That's pretty important arithmetic these days.

Experiments have also been made with so-called "alcohol/water injection systems" . . . which are actually dual carburetor systems in which gasoline is fed to the engine through one line while an alcohol/water mix is fed through the other. Tests run at the Army's Aberdeen Proving Grounds in the 1950's proved that—while such systems lowered gasoline use only slightly (6% on the average)—they allowed gasoline of ten octane numbers *lower* than that required by the test vehicles to be used . . . and resulted in cleaner engines and less cylinder wear, as well.

Separate tests during the same period at the Ohio Agricultural Experiment Station showed that alcohol/water injection provided an increase of 8–10 octane numbers in commercial high-octane gasoline, and up to 15 octane numbers in kerosene and other distillate fuels. This allowed increases in the compression ratios of the test engines which resulted in both power and efficiency increases of 10–20%.

Around the world, experiments with all these systems are being done to lessen dependence upon oil. In Brazil, a blend of 80% alcohol and 20% vegetable oil has reportedly been shown to work efficiently in diesel trucks and tractors. In Germany, a major government-funded program has Volkswagen testing vehicles which burn 15% methanol and 85% low-lead gasoline. This fuel, according to preliminary reports, can be used in existing cars . . .

*Gasoline-burning tractors (as opposed to diesel-powered machines, which are more difficult to convert) can be easily modified to run on alcohol fuel, since the alterations they require are similar to those that would be performed on an automobile.*

*Most 4-cycle stationary engines—such as the unit on this generator—will run smoothly on ethanol after the main metering jet and idle jet (which is being tuned in the photo) are adjusted.*

while meeting all current requirements of performance and reliability.

## HOW TO ADAPT OTHER TYPES OF EQUIPMENT TO BURN ETHYL ALCOHOL

Aside from trucks, automobiles, and tractors, there are some other pieces of equipment that you may wish to operate on alcohol fuel. We'll briefly explain the conversion process for nonvehicle machinery.

### 4-Cycle Engines

Lawn mowers, pumps, and other implements usually use single- or multicylinder utility engines as the prime movers. The conversion process on these engines is uncomplicated, because their carburetors' two metering jets are usually fully adjustable.

First locate the main metering jet (this is the larger one with an oversized head), and open it (counterclockwise) about 1/2 turn. Then find the idle jet and open that screw between 1/2 and 3/4 turn (don't confuse this with the idle speed screw located on the throttle, which should be left alone). That's about all there is to it. Ignition timing is usually fixed on these engines, and can't be adjusted.

A word of caution: When operating the engine on alcohol, keep a close watch on its muffler and exhaust temperature in general. If the exhaust fumes seem excessively hot—or the muffler starts to turn white—open the main jet slightly in order to enrich the mixture.

### 2-Cycle Engines

Chain saws and some industrial engines are 2-cycle models. These will operate well on alcohol if you either [1] open the main jet and idle jet as described in the 4-cycle section (most 2-strokes also have fully adjustable jets), or [2] as an alternative, purchase a carburetor made specifically for alcohol use. These are also adjustable and are usually used for racing purposes when the engines are applied to go-carts or motorcycles. The main difference between the alcohol and gasoline carburetors is that the alcohol-powered models provide for increased fuel flow.

When you operate a 2-stroke engine on alcohol, you'll have to use synthetic oil in the fuel, since petroleum-based oils do not mix well with alcohol. Typical product names are Neo, Klotz, Amzoil, and S.O.F.-7.

## ETHANOL AS A HEATING FUEL

If you don't want to use ethanol for driving to town, you can use it to keep you warm at home.

As a part of our study of the potential of farm-produced ethanol, MOTHER's research staff conducted a series of tests in which an "average" gun-type oil burner was converted to run on this new fuel. The conversion itself isn't difficult . . . in fact, it involves nothing more than a simple adjustment of the unit's air valve (a part which is standard equipment on all oil-burning furnaces). Once this is done, the alcohol-consuming burner produces a nearly invisible blue flame . . . one that not only heats well, but combusts almost completely and leaves hardly any residue.

The oil burner we chose as a "guinea pig" is the type most often used to fire hot water tanks, although the same model could also be used in a space-heating system. The unit was designated Model AF, and manufactured by the Beckett Corporation of Elyria, Ohio. Original specifications called for No. 1 or No. 2 heating oil as a fuel source, and our burner was equipped with a .40-gallon-per-hour feed nozzle, although tips with a firing rate of up to 3 gallons per hour are available for additional heat output.

Our experiments were designed to determine whether we could use a low-strength alcohol and still produce an efficient "burn". In other words, proving that 200-proof (100%) alcohol worked in the burner would be fine, but if the device could run well on a 160-proof mixture (80% ethanol and 20% water) it would not only be more economical, but might provide more heat as well . . . since the steam generated during combustion would improve transfer of heat to the furnace walls.

With this in mind, MOTHER's research team first fired the burner up with No. 2 fuel oil to observe its performance under "normal" use. As expected, the device produced a good amount of heat, but—even with the air intake adjusted to the optimum—the flame was yellow . . . hardly an indication of a "clean" burn.

The next step was to drain the fuel oil from the system and fill our storage tank with 200-proof ethyl alcohol. After opening up the air valve to "lean out" the mixture (pure alcohol needs plenty of oxygen if it's to combust efficiently), the researchers lit the heater again. And, following a bit of fine tuning, the burner threw out a clean blue flame!

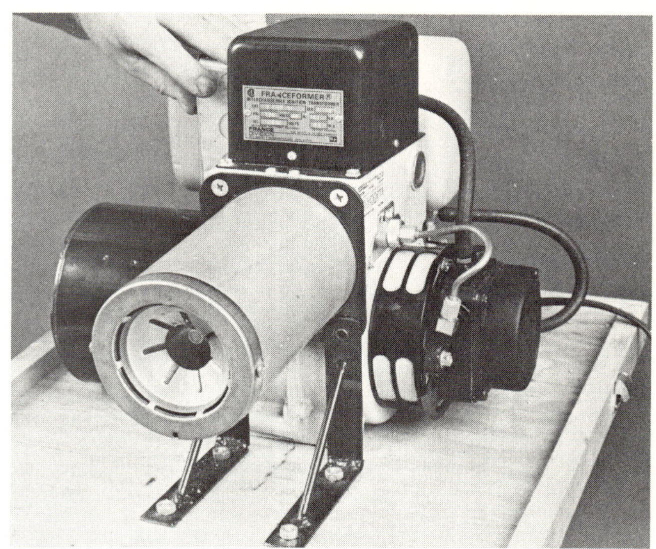


*MOTHER's researchers tested the home-heating capabilities of ethanol, using this "gun-type" oil burner (a unit designed to "fire" a hot water tank, but similar to those used in furnaces).*

*When used with No. 2 fuel oil—as the device was originally designed to be—the burner threw out a good deal of heat (just as we'd expected it to) and a relatively "dirty" yellow flame.*

*On alcohol, the flame was blue and so "clean" as to be almost invisible! We operated the burner on everything from "pure" 200-proof alcohol down to a 75% alcohol/25% water mixture!*

At that point, our shop crew reduced the strength of the alcohol slightly by adding some water (1.6 ounces of water added to 30.4 ounces of pure ethanol yields one quart of 190-proof—or 95%—alcohol) and ran that mixture through the system. Because of this modification to the fuel, the air valve had to be closed a bit before the burner produced the same high-quality, clear blue flame (the oxygen contained in water tends to "aerate" the combustion slightly).

The same efficient burn proved to be possible down to a 150-proof fuel mixture (8 ounces of water and 24 ounces of ethyl alcohol). We did, however, have to continuously decrease the amount of air introduced into the burner as the proof strength dropped . . . to compensate for the oxygen already contained in the diluted alcohol.

Our testers *did* uncover one easily avoidable disadvantage concerning the 150-proof ethanol fuel, however: Because the liquid contained a full 25% water, the pump mechanism on the burner apparently wasn't being fully lubricated and tended to whine while in operation. This problem was solved by installing a "T" fitting in the fuel line (on the suction side of the pump) and plumbing a small storage tank into the "T" . . . with a needle valve incorporated into the line between the two components. The staffers then merely filled the tank with kerosene, adjusted the valve to admit lubricant by the drop, and put the furnace back into operation . . . this time without any noise. The kerosene, of course, burns after it passes through the pump, and—because of the minute quantity involved—doesn't affect the quality of the alcohol's combustion. (NOTE: MOM's researchers used a *new* burner unit for their tests, and the gears in the pump weren't fully "worn in" at the time. It's very likely that a well-used furnace pump wouldn't "complain" at all, though it would probably be a good idea to install the drip-feed lubrication system as a precaution . . . *if* you want to burn your alcohol at 150 proof.)

So, as you can see, the conversion to alcohol doesn't involve *any* technical knowledge or intricate tools, and anyone with an oil-burning furnace can do it! Best of all, besides being potentially more economical than heating oil, alcohol burns with an unbelievably clean flame . . . and the resulting heat output is comparable to that produced by the more conventional fuels.

# 8 LEGALITIES

As we've already mentioned, the distillation of alcohol is rigorously controlled by the government. We've all seen colorful movies about moonshiners and bootleggers and the prison sentences (or other painful fates) which awaited them. Well, allowing a little extra drama for poetic license, the tales are true. The U.S. Bureau of Alcohol, Tobacco, and Firearms (ATF) has as one of its primary responsibilities the destruction of illegal stills and the apprehension of the owners of such equipment. And the law doesn't make any distinctions about what your alcohol is used for when it comes to cracking down on you for operating a still without permission. So it's important that you acquire a permit to operate a still before you build one . . . and it's just as important that you abide by the rules once your distillation apparatus is built.

In 1959, the government made it possible to obtain what it calls an "experimental distilled spirits plant" permit (XDSP). Fortunately, the requirements are less stringent than those for commercial distilleries.

The ATF has supplied the following information about how to make preliminary application for a permit to operate an experimental plant.

*Under Title 27, Code of Federal Regulations, Section 201.64, anyone may establish an experimental distilled spirits plant for experimentation in or development of*

- *[a] Sources of material from which spirits may be produced.*
- *[b] Processes by which spirits may be produced or refined.*
- *[c] Industrial uses of spirits.*

*There is no ATF application form to fill out initially. You simply submit a letter—with certain pertinent information presented in your own words—to your regional office of the Bureau of Alcohol, Tobacco, and Firearms. The following information is required in this letter:*

### Nature and Purpose

*The purpose of the experimental operation: Give a general statement describing what you intend to do.*

### Description of Plant Premises

*Describe the location of the premises where you intend to establish the experimental plant. If you are a farmer, this should include your entire farm . . . to enable you to use the alcohol you produce without removing it from your "plant premises". Include in your description the number of acres involved. Also, describe the buildings used in the production and storage of alcohol (if applicable) and their relative location on your farm. (Drawings and/or survey plats of the property are helpful.)*

### *Description of Production Process and Equipment*

*Describe the production process you intend to use. Then give a descriptive list of the equipment to be used in the process. Describe when and how you intend to denature (make unfit for drinking) the alcohol produced. Also, advise us what you intend to do with the by-products of the process (cattle feed, fertilizer, etc.).*

### *Security*

*Tell us what security measures will be provided for the alcohol produced.*

### *Rate of Production*

*State, in gallons, the amount of alcohol you expect to produce in a 15-day period. Tell us the probable average proof of the finished alcohol.*

### *Duration of Proposed Operations*

*Tell us how long you desire to operate the experimental distilled spirits plant (period not to exceed two years).*

Alcohol produced under the experimental operation may not be sold or given away. If the alcohol is not used on the defined plant premises, it may be subject to a tax or to denaturing regulations.

We suggest that you also contact your state and local authorities for their requirements relating to the production and use of alcohol for experimental purposes.

If you have questions, telephone or write your state or regional ATF office (see the lists of addresses that follow) . . . the people there are your best source of information about the regulations that apply to your particular area of the country, and can advise you of the exact steps to take once you've written your letter.

The office to which you apply will provide whatever forms it may require you to prepare. These may include any of the following that are pertinent to your situation:

| | |
|---|---|
| ATF Form 3A | Indemnity Bond |
| ATF Form 1534 (5000.8) | Power of Attorney |
| ATF Form 1602 | Consent |
| ATF Form 2601 (5110.56) | Distilled Spirits Bond |
| ATF Form 2603 (5110.25) | Application for Operating Permit Under 26 U.S.C. 5171(b) |
| ATF Form 2607 (5110.41) | Registration of Distilled Spirits Plant |
| ATF Form 2625 (5000.9) | Personnel Questionnaire—Alcohol and Tobacco Products |
| ATF Form 4805 (1740.2) | Supplemental Information on Water Quality Considerations—Under 33 U.S.C. 1341(a) |
| ATF Form 4871 (1740.1) | Environmental Information |
| ATF Form 5100.1 | Signing Authority for Corporate Officials |

If you go about the permit procedure correctly, you'll find that the ATF people—at local, state, and national levels—are usually disposed to be very cooperative and helpful. The home fuels movement is growing, and the ATF people realize that the old "moonshine" laws aren't really appropriate for small-scale home fuel distilleries. So the regulators are trying to keep their application requirements at a reasonable level of complexity.

One helpful development involves the bond requirements. Because they feared that folks might use their stills to produce alcohol for other than fuel purposes, the government originally required that a $10,000 bond be posted . . . and since bonding companies aren't familiar with the processes or risks that can be involved in small-scale distilling, it hasn't been easy for the small distiller to be bonded for a reasonable rate.

That's all changed now. A bond is no longer required for small-scale distilleries. (The ATF defines *small-scale* as producing less than 10,000 proof gallons of distilled spirits per calendar year.)

More recently, the Congress of the United States —through its passage of the Crude Oil Windfall Profit Tax Act of 1980—has made it even easier to obtain the permission you'll need to operate an alcohol fuel plant. The new procedure significantly "streamlines" the process of meeting regulations for those persons interested in producing alcohol *solely for fuel purposes*!

Under the new regulations—effective July 1, 1980—an *alcohol fuel producer* (AFP) may produce alcohol fuel for sale, and the permits remain in effect until terminated, revoked, or voluntarily suspended. (The licenses are *non*transferable, and thus are automatically terminated in the event of a lease or sale of either the operations or the corporation which holds the operations.) What's more, *no* bond will be required from plant operators who annually produce fewer than 10,000 proof gallons.

In order to apply for the new AFP permit, just contact your nearest ATF office and ask for form 5110.74. An ATF representative will provide you with all the necessary application guidelines and requirements.

Within 15 days of receiving your application, the regional regulatory administrator will return a written notice of receipt. The notice will include a statement as to whether the application meets all ATF requirements. If it is incomplete, the form will be returned for amendments and/or corrections.

Once it receives a completed application, the ATF has 45 days to issue or deny a permit. If you don't hear from the ATF within the 45-day limit, you may consider the application approved and proceed with the operation of the plant.

(Note also that since July 1, 1980, anyone who applies for an experimental permit—with the intent to make ethanol *fuel*—is automatically considered for an AFP permit.)

# ADDRESSES OF ATF REGIONAL OFFICES

## CENTRAL REGION

Indiana, Kentucky, Michigan, Ohio, West Virginia

Regional Regulatory Administrator
Bureau of Alcohol, Tobacco, and Firearms

550 Main Street
Cincinnati, Ohio 45202

## MID-ATLANTIC REGION

Delaware, District of Columbia, Maryland, New Jersey, Pennsylvania, Virginia

Regional Regulatory Administrator
Bureau of Alcohol, Tobacco, and Firearms

2 Penn Center Plaza, Room 360
Philadelphia, Pennsylvania 19102

## MIDWEST REGION

Illinois, Iowa, Kansas, Minnesota, Missouri, Nebraska, North Dakota, South Dakota, Wisconsin

Regional Regulatory Administrator
Bureau of Alcohol, Tobacco, and Firearms

230 S. Dearborn Street
15th Floor
Chicago, Illinois 60604

## NORTH-ATLANTIC REGION

Connecticut, Maine, Massachusetts, New Hampshire, New York, Rhode Island, Vermont, Puerto Rico, Virgin Islands

Regional Regulatory Administrator
Bureau of Alcohol, Tobacco, and Firearms

6 World Trade Center, 6th Floor
(Mail: P.O. Box 15, Church Street Station)
New York, New York 10008

## SOUTHEAST REGION

Alabama, Florida, Georgia, Mississippi, North Carolina, South Carolina, Tennessee

Regional Regulatory Administrator
Bureau of Alcohol, Tobacco, and Firearms

3835 Northeast Expressway
(Mail: P.O. Box 2994)
Atlanta, Georgia 30301

## SOUTHWEST REGION

Arkansas, Colorado, Louisiana, New Mexico, Oklahoma, Texas, Wyoming

Regional Regulatory Administrator
Bureau of Alcohol, Tobacco, and Firearms

Main Tower, Room 345
1200 Main Street
Dallas, Texas 75202

## WESTERN REGION

Alaska, Arizona, California, Hawaii, Idaho, Montana, Nevada, Oregon, Utah, Washington

Regional Regulatory Administrator
Bureau of Alcohol, Tobacco, and Firearms

525 Market Street
34th Floor
San Francisco, California 94105

# STATE-BY-STATE ACCESS GUIDE TO ALCOHOL FUEL LAWS

The following list will help those who are interested in establishing their own alcohol fuel plants to comply with the individual states' requirements in this area. The agency noted under each of the 50 state headings—plus the District of Columbia—can provide official information regarding the production of ethanol for fuel, including permit and bonding regulations . . . sales, use, excise, or gross receipts tax requirements . . . and environmental, fire, health, labor, and public safety laws. The states' representatives can also make a potential fuel producer aware of available grants or loans, and property or income tax reductions, that he or she might have otherwise overlooked.

**ALABAMA**
Fred Braswell, Program Coordinator, Alabama Energy Management Board, 3734 Atlanta Hwy., Montgomery, Ala. 36130, 205/832-5010

**ALASKA**
Paula Wellen, Community Information Center, Fairbanks North Star Borough, 520 Fifth Ave., Fairbanks, Alaska 99707, 907/452-4761

**ARIZONA**
Richard Wetzel, Arizona Office of Economic Planning and Development, 1700 W. Washington, Executive Tower 505, Phoenix, Ariz. 85007, 602/255-5705

**ARKANSAS**
Alford Drinkwater, Biomass/Resource Recovery Coordinator, Arkansas Department of Energy, 3000 Kavanaugh, Little Rock, Ark. 72205, 501/371-1370

**CALIFORNIA**
Manuel R. Espinoza, Chief of Business Practices, California Department of Alcoholic Beverage Control, 1215 O St., Sacramento, Calif. 95814, 916/445-4687

**COLORADO**
Bob Merten, Department of Agriculture, 525 Sherman St., 4th Floor, Denver, Colo. 80203, 303/839-3218

**CONNECTICUT**
Joe Belanger, Director, Energy Research and Policy, 81 Washington St., Hartford, Conn. 06115, 203/566-5898

**DELAWARE**
Dan Anstine, Delaware Energy Office, P.O. Box 1401, 114 W. Water St., Dover, Del. 19901, 302/736-5647

**FLORIDA**
Will Kirksey, Governor's Energy Office, 301 Bryant Bldg., Tallahassee, Fla. 32301, 904/488-6146

**GEORGIA**
Rob Harvey, Office of Energy Resources, 270 Washington St. S.W., Room 615, Atlanta, Ga. 30334, 404/656-5176

**HAWAII**
Takeshi Yoshihara, Hawaii Representative, Department of Energy, 4322 Prince Kuhio Federal Bldg., Honolulu, Hawaii 96850, 808/546-3730

**IDAHO**
Gail Dameworth, Research Analyst, Idaho Office of Energy, State House Mail, Boise, Idaho 83720, 208/334-3800

**ILLINOIS**
Nicholas P. Hall, Manager, Alcohol Fuels Program, Resource Development Division, Room 300, 325 W. Adams St., Springfield, Ill. 62706, 217/785-2800

**INDIANA**
Mary Failey, Department of Commerce, 440 North Meriden, Indianapolis, Ind. 46201, 317/232-8954

**IOWA**
Doug Getter, Director, Administrative Services, Iowa Development Commission, 250 Jewett Bldg., Des Moines, Iowa 50309, 515/281-3251

**KANSAS**
Randy Noon, State Energy Office, 214 W. 6th St., Topeka, Kan. 66603, 913/296-2496

**KENTUCKY**
Bruce Sauer, Kentucky Department of Energy, P.O. Box 11888, Lexington, Ky. 40578, 606/252-5535

**LOUISIANA**
Thomas C. Landrum, Director, Department of Natural Resources, Division of Research and Development, P.O. Box 44156, Capitol Station, Baton Rouge, La. 70804, 504/342-4594

**MAINE**
Nancy Holmes, Office of Energy Resources, 55 Capitol St., Augusta, Maine 04330, 207/289-3811

**MARYLAND**
Marvin Bond, Office of Public Affairs, P.O. Box 466, Annapolis, Md. 21404, 301/269-3885

**MASSACHUSETTS**
Chris Hansen, Executive Office of Energy Resources, 73 Tremont St., Boston, Mass. 02108, 617/727-1990

**MICHIGAN**
Randy Harmson, Energy Coordinator, Marketing and International Trade Division, Michigan Department of Agriculture, P.O. Box 30017, Lansing, Mich. 48909, 517/373-1054

**MINNESOTA**
Dennis M. Devereaux, Alternative Energy Projects, Minnesota Energy Agency, 980 American Center Bldg., 150 East Kellogg Blvd., St. Paul, Minn. 55101, 612/296-9078

**MISSISSIPPI**
Robert Smira, Conservation Director, Office of Energy, Department of Natural Resources, 300 Watkins Bldg., 510 George St., Jackson, Miss. 39202, 601/961-4403

**MISSOURI**
Deborah Goldhammer, Manager of Program Development, Division of Energy, P.O. Box 176, Jefferson City, Mo. 65102, 314/751-4000

**MONTANA**
Georgia Brensbal, Program Engineer, Renewable Energy Bureau, Department of Natural Resources and Conservation, 32 South Ewing, Helena, Mont. 59601, 406/449-4624

**NEBRASKA**
Steve Sorum, Administrator, Agricultural Products Industrial Utilization Committee, 301 Centennial Mall, South 3rd Floor, Lincoln, Neb. 68509, 402/471-2941

**NEVADA**
Mr. Kelly Jackson, Nevada Department of Energy, 400 W. King St., Room 106, Carson City, Nev. 89710, 702/885-5157

**NEW HAMPSHIRE**
Tina Oleson, Government's Council on Energy, 1/2 Beacon St., Concord, N.H. 03310, 603/271-2771

**NEW JERSEY**
Louis Jarecki, Office of Alternate Technology, New Jersey Department of Energy, 101 Commerce St., Newark, N.J. 07102, 201/648-6293

**NEW MEXICO**
Charles P. Wood, Assistant Director, Energy Resource and Development Division, Energy and Minerals Department, P.O. Box 2770, Santa Fe, N.M. 87501, 505/827-2471

**NEW YORK**
Mark Bagdon, Bureau of Resource Development, New York State Energy Office, Agency Building 2, Rockefeller Plaza, Albany, N.Y. 12210, 518/474-7875

**NORTH CAROLINA**
John Manuel, Energy Division, Department of Commerce, P.O. Box 25249, Raleigh, N.C. 27611, 919/733-4493

**NORTH DAKOTA**
John G. Conrad, State Solar Officer, Energy Management and Conservation Programs, Federal Aid Coordinator Office, State Capitol, Bismarck, N.D. 58505, 701/224-2250

**OHIO**
Bob Yaekle, Ohio Legislative Service Commission, State House, 5th Floor, Columbus, Ohio 42315, 614/466-4573

**OKLAHOMA**
Rex Privett, State Gasohol Coordinator, State Department of Energy, 4400 N. Lincoln Blvd., Suite 35, Oklahoma City, Okla. 73105, 405/427-3829

**OREGON**
Richard L. Durham, Oregon Department of Energy, 102 Labor and Industries Bldg., Salem, Ore. 97310, 503/378-4998

**PENNSYLVANIA**
Research Analyst, Minority Research Unit, House of Representatives, Box 250, Harrisburg, Pa. 17120, 717/783-1572

**RHODE ISLAND**
Shelly Greenfield, Governor's Energy Office, 80 Dean St., Providence, R.I. 02903, 401/277-3374

**SOUTH CAROLINA**
Cathy Twilley, Administrative Assistant, Governor's Division of Energy Resources, 1122 Lady St., Columbia, S.C. 29201, 803/758-8110

**SOUTH DAKOTA**
Verne Brakke, Office of Energy Policy, Pierre, S.D. 57501, 605/773-3603

**TENNESSEE**
Margot Myrick, Research Analyst, Legislative Plaza, Suite 2, Nashville, Tenn. 37219, 615/741-2354

**TEXAS**
Bob Avant, Coordinator of Biomass and Wind Programs, Texas Energy and Natural Resources Advisory Council, 411 W. 13th St., Suite 900, Austin, Tex. 78701, 512/475-5588

**UTAH**
Jim Bryne, Deputy Director, Utah Energy Office, 231 E. 400 South, Salt Lake City, Utah 84111, 801/533-5424

**VERMONT**
Larry Ogden, State Energy Office, State Office Bldg., Montpelier, Vt. 05602, 802/828-2393

**VIRGINIA**
Irl Smith, State Director of Alcohol Fuels, Virginia Department of Agriculture & Consumer Services, 203 N. Governors St., Richmond, Va. 23219, 804/786-3519

**WASHINGTON**
Paul Juhasz, State Energy Office, 400 E. Union St., Olympia, Wash. 98504, 206/754-0700

**WEST VIRGINIA**
Rebecca Scott, West Virginia Fuel and Energy Office, 1262½ Greenbriar St., Charleston, W.Va. 25311, 304/348-8860

**WISCONSIN**
George Plaza, Biomass Analyst, Division of State Energy, 101 S. Webster St., Madison, Wis. 53702, 608/266-0985

**WYOMING**
Butch Keadle, Fuel Allocation Specialist, Energy Conservation Office, Capitol Hill Office Bldg., 25th and Pioneer, Cheyenne, Wyo. 80002, 307/777-7284

**DISTRICT OF COLUMBIA**
Program Director, D.C. Energy Unit, Suite 200, 1420 New York Ave. N.W., Washington, D.C. 20005, 202/727-1825

# APPENDIX

# A HOT-OIL FURNACE/6" STILL COMBINATION

Put simply, MOTHER's still heater is little more than a woodstove surrounded by an oil-filled chamber that's specifically designed to expose a maximum amount of surface area to the heat source. This optimum thermal transfer is accomplished not only by *encircling* the cylinder-shaped firebox with the viscous liquid (we use Exxon's Caloria HT-43, which has been formulated to handle high temperatures), but by cleverly routing the stove's smoke (and hence its normally wasted heat) directly *through* the oil reservoir via a series of 1-1/2"-diameter tubes.

Now some of you may question our use of oil—rather than free-for-the-taking water—as a thermal transfer medium . . . but there are several good reasons why the choice was made:

[1] Oil—like water—can hold heat for an extended period of time.

[2] The boiling temperature of oil is more than twice that of water . . . therefore a lot more heat can be fed into it—at atmospheric pressure—before a molecular change occurs.

[3] Oil is noncorrosive, and thus its use results in a longer life span for the furnace *and* for the plumbing connected to it.

Another important point to remember (for reasons of both safety and legality) is that, although MOTHER's oil-heating furnace may *look* like a boiler, it is definitely not one . . . since it's open to the air and operates at atmospheric pressure. This means—in essence—that those laws, inspections, and other requirements that apply to steam boilers do *not* pertain to this heater. (Incidentally, because any water in the oil will be vaporized—and vented—long before the heat storage medium reaches its operating temperature, neither condensation nor the accidental introduction of water into the oil reservoir will pose a problem.)

*MOTHER's hot-oil furnace is a woodstove surrounded by an oil-filled chamber to provide thermal transfer. The heating unit can be teamed with a still for fermentation and distillation.*

## ONE OF MANY APPLICATIONS

Naturally, the furnace described here could be used—in conjunction with a heat exchanger—for space heating, water heating, evaporation-based cooling or refrigeration, or for any of a *number* of other purposes. But since, THE MOTHER EARTH NEWS® alcohol researchers needed a compact energy source to provide heat for fermentation and

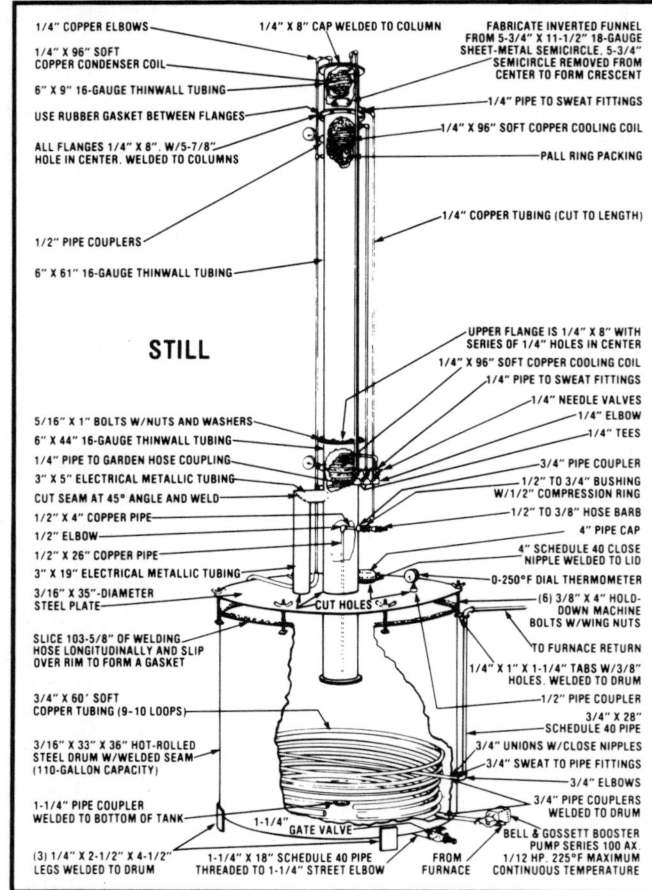


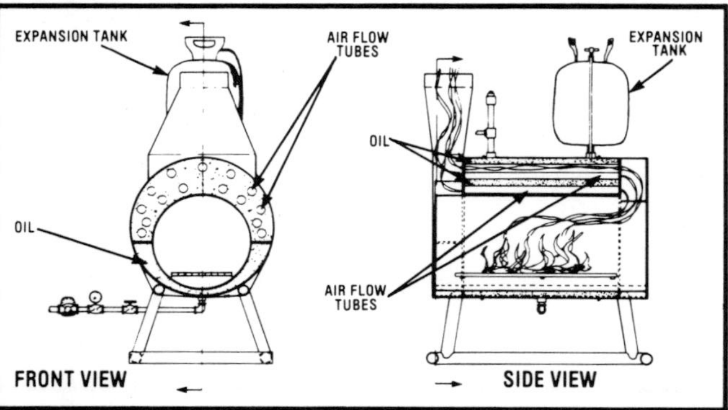


distillation, we've chosen to feature both the furnace *and* a modified distillery together . . . so that people who choose to do so can build the two projects so that they'll work in tandem.

The accompanying drawings are probably all but self-explanatory, but we *do* have a few tips that will make the task of fabricating your equipment somewhat easier. In most cases, both of the furnace's "drums" will have to be formed—from flat plates—at your local steel supply house (unless you just happen to have on hand two tanks of about the required sizes).

By the same token, the perforated end plates that seal the extremities of the oil chamber should be drilled simultaneously, to ensure that the holes line up perfectly. (As an alternative to cutting the openings yourself, you'll find that almost any sheet metal shop will stamp the plates, accurately, at a reasonable cost.) The chimney stack may also be awkward to piece together, since the fitting involves joining a circular collar to a four-sided opening . . . so this component, too, might best be made by a professional.

(As far as the still is concerned, just about any steel vat will suffice for the mash tank . . . as long as it's sealed effectively at its lid. We made the entire top removable, for easy cleaning and to facilitate installation of the heat exchanger coils. The three condenser coils *within* the column should also be carefully formed . . . using the tube roller detailed in our illustration. Simply insert one end of your eight-foot section of tubing into the indicated notch and roll it around the center hub—keeping it flush with the base plate—until your spiral's outer edge reaches the dotted line. Then repeat the procedure, starting at the *uncoiled* end of the pipe. When both halves are formed, merely pull the ends apart, accordion-style, till the coiled condenser is about eight inches long.)

## AND IT WORKS LIKE A CHARM!

We're happy to report that the furnace design has proved itself to be sound, and that the unit is impressively efficient. Flue temperatures average *below* 400 °F, which means that a good deal of heat *is* going into the oil rather than up the chimney. And, as long as the welds aren't defective and oil temperatures never exceed the liquid's flash point (435 °F), the apparatus is perfectly safe.

So far, we've been nothing but pleased with the

way our little backyard *non*boiler works . . . but we've got a few ideas up our collective sleeve which —we hope—will increase the practicality (and/or lower the operating cost) of the furnace. These include burying the container in a sand pit to minimize loss of heat, installing an *oil-burner* unit—on an interchangeable door—to provide maintenance heat when the operator has to be away, and even experimenting with *corn* as the sole source of raw material to feed the furnace/still combination. (We hope the kernels will provide oil for heat storage *plus* starch and sugar for alcohol production, while the stalks and cobs serve as fuel for the firebox!)

NO. 6 X 1-1/4" WOOD SCREWS
3/4" X 2" WOODEN DISK GLUED AND FASTENED TO BASE
1/4" X 4" BOLT W/WING NUT (SECURE TO WORK TABLE)
MARK LIMIT LINE AT 2-3/4" RADIUS FROM CENTER
COIL-ROLLING TOOL
CUT 3/8" X 1-1/4" NOTCH IN SIDE OF SMALL DISK
DRIVE FINISHING NAIL THROUGH OUTER EDGE OF NOTCH
3/4" X 8" WOODEN DISK

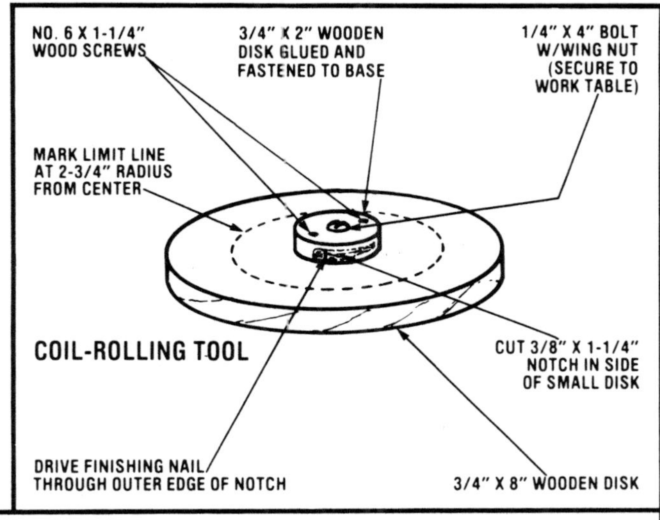

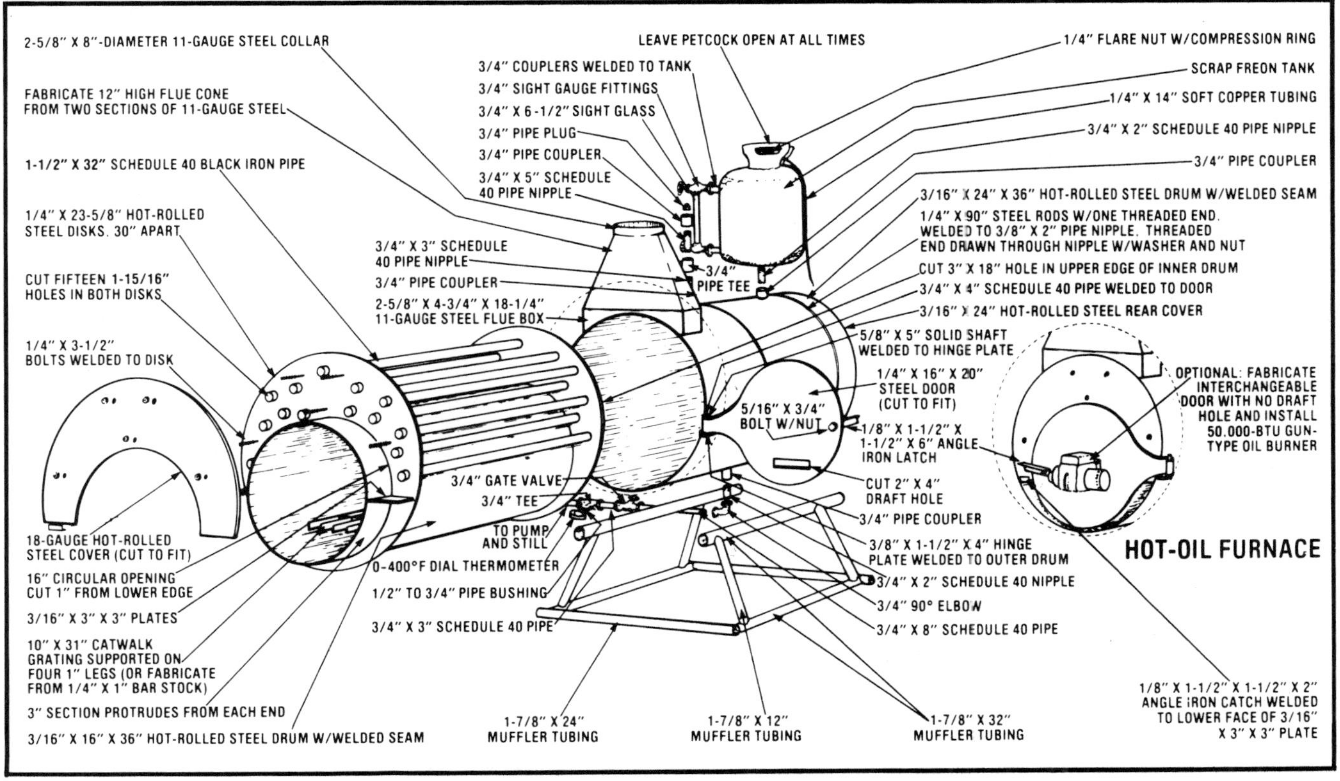

2-5/8" X 8"-DIAMETER 11-GAUGE STEEL COLLAR
FABRICATE 12" HIGH FLUE CONE FROM TWO SECTIONS OF 11-GAUGE STEEL
1-1/2" X 32" SCHEDULE 40 BLACK IRON PIPE
1/4" X 23-5/8" HOT-ROLLED STEEL DISKS, 30" APART
CUT FIFTEEN 1-15/16" HOLES IN BOTH DISKS
1/4" X 3-1/2" BOLTS WELDED TO DISK
18-GAUGE HOT-ROLLED STEEL COVER (CUT TO FIT)
16" CIRCULAR OPENING CUT 1" FROM LOWER EDGE
3/16" X 3" X 3" PLATES
10" X 31" CATWALK GRATING SUPPORTED ON FOUR 1" LEGS (OR FABRICATE FROM 1/4" X 1" BAR STOCK)
3" SECTION PROTRUDES FROM EACH END
3/16" X 16" X 36" HOT-ROLLED STEEL DRUM W/WELDED SEAM
LEAVE PETCOCK OPEN AT ALL TIMES
3/4" COUPLERS WELDED TO TANK
3/4" SIGHT GAUGE FITTINGS
3/4" X 6-1/2" SIGHT GLASS
3/4" PIPE PLUG
3/4" PIPE COUPLER
3/4" X 5" SCHEDULE 40 PIPE NIPPLE
3/4" X 3" SCHEDULE 40 PIPE NIPPLE
3/4" PIPE COUPLER
2-5/8" X 4-3/4" X 18-1/4" 11-GAUGE STEEL FLUE BOX
3/4" PIPE TEE
3/4" GATE VALVE
3/4" TEE
TO PUMP AND STILL
0-400°F DIAL THERMOMETER
1/2" TO 3/4" PIPE BUSHING
3/4" X 3" SCHEDULE 40 PIPE
1-7/8" X 24" MUFFLER TUBING
1-7/8" X 12" MUFFLER TUBING
1-7/8" X 32" MUFFLER TUBING
1/4" FLARE NUT W/COMPRESSION RING
SCRAP FREON TANK
1/4" X 14" SOFT COPPER TUBING
3/4" X 2" SCHEDULE 40 PIPE NIPPLE
3/4" PIPE COUPLER
3/16" X 24" X 36" HOT-ROLLED STEEL DRUM W/WELDED SEAM
1/4" X 90" STEEL RODS W/ONE THREADED END, WELDED TO 3/8" X 2" PIPE NIPPLE, THREADED END DRAWN THROUGH NIPPLE W/WASHER AND NUT
CUT 3" X 18" HOLE IN UPPER EDGE OF INNER DRUM
3/4" X 4" SCHEDULE 40 PIPE WELDED TO DOOR
3/16" X 24" HOT-ROLLED STEEL REAR COVER
5/8" X 5" SOLID SHAFT WELDED TO HINGE PLATE
1/4" X 16" X 20" STEEL DOOR (CUT TO FIT)
5/16" X 3/4" BOLT W/NUT
1/8" X 1-1/2" X 1-1/2" X 6" ANGLE IRON LATCH
CUT 2" X 4" DRAFT HOLE
3/4" PIPE COUPLER
3/8" X 1-1/2" X 4" HINGE PLATE WELDED TO OUTER DRUM
3/4" X 2" SCHEDULE 40 NIPPLE
3/4" 90° ELBOW
3/4" X 8" SCHEDULE 40 PIPE
OPTIONAL: FABRICATE INTERCHANGEABLE DOOR WITH NO DRAFT HOLE AND INSTALL 50,000-BTU GUN-TYPE OIL BURNER
HOT-OIL FURNACE
1/8" X 1-1/2" X 1-1/2" X 2" ANGLE IRON CATCH WELDED TO LOWER FACE OF 3/16" X 3" X 3" PLATE

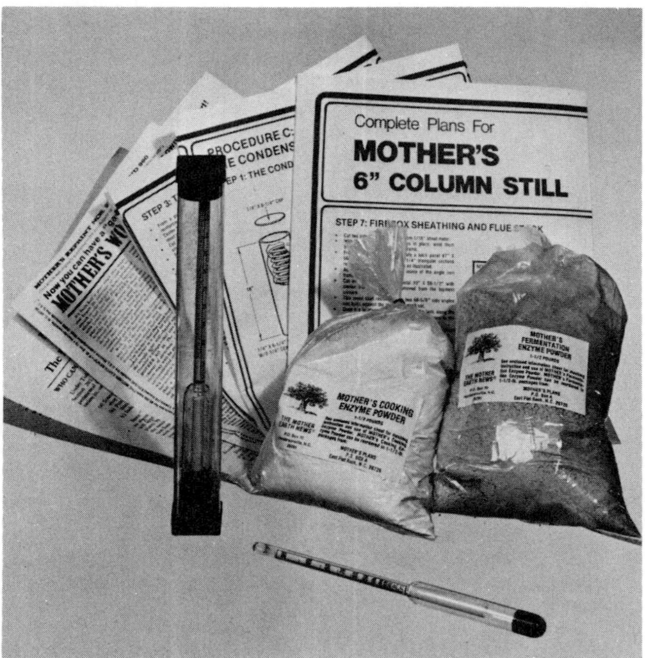


*MOTHER's equipment for homestead alcohol fuel production . . . still plans, cooking and fermentation enzymes, a spirit hydrometer, and a saccharometer (or triple-scale hydrometer).*

# EQUIPMENT

Of course, you can't build *everything* you'll need to make your own ethanol. If you use enzymes in your mash (and we recommend that you do), you'll need to purchase them somewhere. You'll also need a hydrometer and other materials.

MOTHER has put together a kit containing fully illustrated, detailed construction plans for either her three- *or* her six-inch column still. Included in the package are supplies of MOTHER's Mash Cooking Enzyme Powder (an alpha-amylase) and Fermenting Powder (which is a denaturant, glucoamylase, and yeast combination)—enough of each to process *at least* 20 bushels of corn—plus a hydrometer, a saccharometer, and an instruction sheet. The kit retails for $35. The kit may also include a copy of THE MOTHER EARTH NEWS® *Alcohol Fuel Handbook*, in which case it retails for $45. For each kit, be sure to specify whether three- or six-inch column still plans should be included. Kits are available from MOTHER'S PLANS, P.O. Box A, East Flat Rock, North Carolina 28726. The items in the package can also be purchased separately, as follows:

| Item | No. | Price |
|---|---|---|
| MOTHER's Still Plans<br>3″ plans<br>6″ plans | <br>84016<br>84027 | $15.00 postpaid<br>(be sure to specify whether you want the three- or the six-inch column design) |
| 1-1/2 lb. MOTHER's Alcohol Fuel Mash Cooking Enzyme | 6053 | $8.40, plus $1.00 |
| 1-1/2 lb. MOTHER's Alcohol Fuel Mash Fermentation Enzyme With Yeast | 6054 | $8.90, plus $1.00 |
| Hydrometer (spirit) | 6050 | $4.95, plus $1.00 |
| Saccharometer (mash) | 6051 | $5.50, plus $1.00 |

(NOTE: Orders totaling $10.00 or more are shipped postpaid.)

## PRODUCT SOURCE LIST

### LABORATORY SUPPLIES AND EQUIPMENT

Rascher & Betzold, Inc.
5410 N. Damen Avenue
Chicago, Illinois 60625
Phone: 312/275-7300

### CUSTOM-MADE STILL PLATES

Arbor Sales Co. Ltd.
P.O. Box 6
Des Moines, Iowa 50301
Phone: 515/279-8091

### SIMPLE DISTILLATION STILL (80-GALLON BATCH)

Farm Energy Systems
P.O. Box 2573
Pocatello, Idaho 83201

### ENZYMES

Alltech, Inc.
271 Gold Rush Road
Lexington, Kentucky 40503
Phone: 606/276-3414

Aquaterra Biochemicals Corp.
P.O. Box 496
Lancaster, Texas 75146
Phone: 214/227-6960
Glucoamylase only

BIOCON (U.S.), Inc.
261 Midland Avenue
Lexington, Kentucky 40507
Phone: 606/254-0517
Minimum order: 4-kilo box (8.8 lb.)

C.B. Fermentation Industries, Inc.
One N. Broadway
Des Plaines, Illinois 60016

Enzeco
Phone: 212/736-1580

Miles Laboratories (Marshall Division)
1127 Myrtle Street
Elkhart, Indiana 46515
Phone: 219/262-7176
Minimum order: 5 gallons

Novo Laboratories, Inc.
59 Danbury Road
Wilton, Connecticut 06897
Minimum order: 55 gallons

Rohm and Haas Company
Independence Mall West
Philadelphia, Pennsylvania 19105
Phone: 215/592-3000

### YEAST

BIOCON (U.S.), Inc.
261 Midland Avenue
Lexington, Kentucky 40507
Phone: 606/254-0517
Minimum order: 4-kilo box (8.8 lb.)

### PACKING

Maurice A. Knight
Box 109
Akron, Ohio 44309

Metex Inc.
308 Talmadge Road
Edison, New Jersey 08817

Norton Chemical Company
Dept. TR-2
12 East Avenue
Tallmadge, Ohio 44278

# MASHING PROCEDURE FOR DIFFERENT RAW MATERIALS

| RAW MATERIAL | PREPARATION | ADDITIVES (ENZYMES) | PREBOIL |
|---|---|---|---|
| **STARCHES** | | | |
| WHEAT, CORN, RYE, BARLEY, MILO, RICE, CATTAILS | Grind to a fine meal using a 3/16″ screen on a hammermill; add 30 gal. water per bushel. | Add 3 spoons mash cooking powder[1] per bushel. | Raise temp. to 170°F for 15 min.; agitate vigorously. |
| PASTRY WASTE | Break apart, do not grind; add 30 gal. water per 55 lb. | As above. | As above. |
| POTATOES, CASSAVA (MANIOC), TARO | Slice, crush, or break apart; add 10 gal. water per 100 lb., or as little water as possible. | Add 5 spoons mash cooking powder[1] per 100 lb. | None. |
| **SUGARS** | | | |
| SUGAR BEETS, MANGELWURZELS, ARTICHOKE TUBERS | Slice or crush; add 10 gal. water per 100 lb., or as little as possible. | Acid may be added to beets to reach pH 5.0. | None. |
| SWEET SORGHUM, CANE, ARTICHOKE STALKS | Squeeze out juice. | None. | Raise temp. to 180°F for 10 min. to sterilize. |
| MOLASSES, SUGAR PRODUCTS | None. | Molasses from beets may need neutralization with acid. | If necessary, sterilize as above. |
| CHEESE WHEY | None. | None. | None. |
| **CELLULOSE** | Chop straw or soft material. Wood must be fine sawdust or treated with 400°F steam for 2 hr. | Add a 1% caustic solution; hold at 140°F for 3 hr. to separate lignin. | Draw off lignin; neutralize. |

| COOK | COOL DOWN | CULTURE | COMMENTS |
|---|---|---|---|
| Hold at rapid rolling boil for 30 min. | Cool with coil to 170°F; add 3 spoons mash cooking powder[1]; agitate for 30–60 min. | Reduce temp. to 90°F; add 6 spoons mash fermenting powder[1]; agitate for 10 min.; cover. | Results: 9% alcohol. Wheat, rye, and barley may cause foaming: Use Low-Foam[2] or mix with cornmeal. |
| As above. | As above. | As above. | As above, plus remove oil (if content is high) before fermentation. |
| Raise temp. to 180°F for 30 min.; agitate vigorously. | None. | Reduce temp. to 90°F; add 10 spoons mash fermenting powder[1]; agitate 10 min.; cover. | Results: 9% alcohol. |
| Raise temp. to 190°F for 20 min.; agitate. | None. | Reduce temp. to 90°F; add yeast; agitate 10 min.; cover. | Results: 7% alcohol. Beets may require some molasses yeast food[2]. |
| None. | None. | Reduce temp. to 90°F; add water to make 18% sugar; add yeast; agitate 10 min.; cover. | Results: 9% alcohol. Molasses yeast food[2] may be added to increase yield. |
| None. | None. | As above. | Results: 9% alcohol. Use molasses yeast food[2] to insure proper yield. High NaCl content may interfere with fermentation. |
| Raise temp. to 210°F for 10 min. to sterilize. | Separate protein with $NH_4OH$; adjust pH to 5.0. | Reduce temp. to 90°F; add *Kluyveromyces fragilis* or *Torula cremoris* yeast. Fermentation takes only 12 hr. | Results: 3% alcohol. Aeration may increase yield. Whey may be used as liquid with corn, but lactase must be added for a conversion. |
| Cook at 140°F for 4 hr. in 1% solution of Biocellulase[2]. | Remove sugar liquid. | Reduce temp. to 90°F; add brewer's yeast; agitate for 10 min.; cover. | Results: 2.5% alcohol. Acid hydrolysis is an alternative but expensive method. |

[1]See the Appendix section on "Equipment", pages 106–107, for information on product availability

[2]Available from BIOCON, Inc., 261 Midland Ave., Lexington, Ky. 40507

# TEMPERATURE CONVERSION TABLE

| °C | °F | °C | °F | °C | °F | °C | °F | °C | °F | °C | °F |
|---|---|---|---|---|---|---|---|---|---|---|---|
| 0 | 32 | 15.56 | 60 | 32.22 | 90 | 48.89 | 120 | 65.56 | 150 | 82.22 | 180 |
| 0.56 | 33 | 16.11 | 61 | 32.78 | 91 | 49.44 | 121 | 66.11 | 151 | 82.78 | 181 |
| 1.11 | 34 | 16.67 | 62 | 33.33 | 92 | 50.00 | 122 | 66.67 | 152 | 83.33 | 182 |
| 1.67 | 35 | 17.22 | 63 | 33.89 | 93 | 50.56 | 123 | 67.22 | 153 | 83.89 | 183 |
| 2.22 | 36 | 17.78 | 64 | 34.44 | 94 | 51.11 | 124 | 67.78 | 154 | 84.44 | 184 |
| 2.78 | 37 | 18.33 | 65 | 35.00 | 95 | 51.67 | 125 | 68.33 | 155 | 85.00 | 185 |
| 3.33 | 38 | 18.89 | 66 | 35.56 | 96 | 52.22 | 126 | 68.89 | 156 | 85.56 | 186 |
| 3.89 | 39 | 19.44 | 67 | 36.11 | 97 | 52.78 | 127 | 69.44 | 157 | 86.11 | 187 |
| 4.44 | 40 | 20.00 | 68 | 36.67 | 98 | 53.33 | 128 | 70.00 | 158 | 86.67 | 188 |
| 5.00 | 41 | 20.56 | 69 | 37.22 | 99 | 53.89 | 129 | 70.56 | 159 | 87.22 | 189 |
| 5.56 | 42 | 21.11 | 70 | 37.78 | 100 | 54.44 | 130 | 71.11 | 160 | 87.78 | 190 |
| 6.11 | 43 | 21.67 | 71 | 38.33 | 101 | 55.00 | 131 | 71.67 | 161 | 88.33 | 191 |
| 6.67 | 44 | 22.22 | 72 | 38.89 | 102 | 55.56 | 132 | 72.22 | 162 | 88.89 | 192 |
| 7.22 | 45 | 22.78 | 73 | 39.44 | 103 | 56.11 | 133 | 72.78 | 163 | 89.44 | 193 |
| 7.78 | 46 | 23.33 | 74 | 40.00 | 104 | 56.67 | 134 | 73.33 | 164 | 90.00 | 194 |
| 8.33 | 47 | 23.89 | 75 | 40.56 | 105 | 57.22 | 135 | 73.89 | 165 | 90.56 | 195 |
| 8.89 | 48 | 24.44 | 76 | 41.11 | 106 | 57.78 | 136 | 74.44 | 166 | 91.11 | 196 |
| 9.44 | 49 | 25.00 | 77 | 41.67 | 107 | 58.33 | 137 | 75.00 | 167 | 91.67 | 197 |
| 10.00 | 50 | 25.56 | 78 | 42.22 | 108 | 58.89 | 138 | 75.56 | 168 | 92.22 | 198 |
| 10.56 | 51 | 26.11 | 79 | 42.78 | 109 | 59.44 | 139 | 76.11 | 169 | 92.78 | 199 |
| 11.11 | 52 | 26.67 | 80 | 43.33 | 110 | 60.00 | 140 | 76.67 | 170 | 93.33 | 200 |
| 11.67 | 53 | 27.22 | 81 | 43.89 | 111 | 60.56 | 141 | 77.22 | 171 | 93.89 | 201 |
| 12.22 | 54 | 27.78 | 82 | 44.44 | 112 | 61.11 | 142 | 77.78 | 172 | 94.44 | 202 |
| 12.78 | 55 | 28.33 | 83 | 45.00 | 113 | 61.67 | 143 | 78.33 | 173 | 95.00 | 203 |
| 13.33 | 56 | 28.89 | 84 | 45.56 | 114 | 62.22 | 144 | 78.89 | 174 | 95.56 | 204 |
| 13.89 | 57 | 29.44 | 85 | 46.11 | 115 | 62.78 | 145 | 79.44 | 175 | 96.11 | 205 |
| 14.44 | 58 | 30.00 | 86 | 46.67 | 116 | 63.33 | 146 | 80.00 | 176 | 96.67 | 206 |
| 15.00 | 59 | 30.56 | 87 | 47.22 | 117 | 63.89 | 147 | 80.56 | 177 | 97.22 | 207 |
| | | 31.11 | 88 | 47.78 | 118 | 64.44 | 148 | 81.11 | 178 | 97.78 | 208 |
| | | 31.67 | 89 | 48.33 | 119 | 65.00 | 149 | 81.67 | 179 | 98.33 | 209 |
| | | | | | | | | | | 98.89 | 210 |
| | | | | | | | | | | 99.44 | 211 |
| | | | | | | | | | | 100.00 | 212 |

**TEMPERATURES OF SPECIAL INTEREST**

**Pure (100%) alcohol boils at 173 °F (78.33 °C)**

**100-proof (50%) alcohol boils at 185 °F (85 °C)**

# Alcohol Production Record Sheet ● ____________

| MASHING | | | DISTILLING | | | NOTES*** |
|---|---|---|---|---|---|---|
| DATE | AMOUNT AND KIND OF MATERIAL* | GAL. OF WATER | DATE | AMOUNT OF ALCOHOL** | | |
| | | | | HIGH-PROOF | LOW-PROOF | |
| | | | | | | |
| | | | | | | |
| | | | | | | |
| | | | | | | |
| | | | | | | |
| | | | | | | |
| | | | | | | |
| | | | | | | |
| | | | | | | |
| | | | | | | |
| | | | | | | |
| | | | | | | |
| | | | | | | |
| | | | | | | |
| | | | | | | |
| | | | | | | |

*EXAMPLE: GROUND CORN/56 LB.

** STATE AMOUNTS IN GALLONS. EXAMPLE: (2.0/180) (2.0/70)

*** EXAMPLE: USED MOTHER'S THREE-STEP MASHING RECIPE

# Alcohol Production Record Sheet ● ____________

| MASHING | | | DISTILLING | | | NOTES*** |
|---|---|---|---|---|---|---|
| DATE | AMOUNT AND KIND OF MATERIAL* | GAL. OF WATER | DATE | AMOUNT OF ALCOHOL** | | |
| | | | | HIGH-PROOF | LOW-PROOF | |
| | | | | | | |
| | | | | | | |
| | | | | | | |
| | | | | | | |
| | | | | | | |
| | | | | | | |
| | | | | | | |
| | | | | | | |
| | | | | | | |
| | | | | | | |
| | | | | | | |
| | | | | | | |
| | | | | | | |
| | | | | | | |
| | | | | | | |
| | | | | | | |

*EXAMPLE: GROUND CORN/56 LB.

** STATE AMOUNTS IN GALLONS.
EXAMPLE: (2.0/180) (2.0/70)

*** EXAMPLE: USED MOTHER'S THREE-STEP MASHING RECIPE

# Alcohol Denature Record Sheet ● ____________________

| DATE | GALLONS OF ALCOHOL | AMOUNTS OF DENATURANTS * | | | GOV'T FORMULA # | NOTES |
|---|---|---|---|---|---|---|
| | | GASOLINE | | | | |
| | | | | | | |
| | | | | | | |
| | | | | | | |
| | | | | | | |
| | | | | | | |
| | | | | | | |
| | | | | | | |
| | | | | | | |
| | | | | | | |
| | | | | | | |
| | | | | | | |
| | | | | | | |
| | | | | | | |
| | | | | | | |
| | | | | | | |
| | | | | | | |
| | | | | | | |

* LIST DENATURANTS BY NAME AND AMOUNTS IN GALLONS.
EXAMPLE: (KETONE/0.125) (GASOLINE/1.0)

## Alcohol Disposal Record Sheet ● ______________________

| DATE | PROOF | DENATURED YES/NO | GAL. | MANNER OF DISPOSAL OR USE OF ALCOHOL* |
|---|---|---|---|---|
| | | | | |
| | | | | |
| | | | | |
| | | | | |
| | | | | |
| | | | | |
| | | | | |
| | | | | |
| | | | | |
| | | | | |
| | | | | |
| | | | | |
| | | | | |
| | | | | |
| | | | | |
| | | | | |
| | | | | |

* EXAMPLE: FUEL FOR FARM TRACTOR/FIELD WORK

# RECOMMENDED READING

*Alcohols and Hydrocarbons as Motor Fuels.* Warrendale, Pennsylvania: Society of Automotive Engineers, 1964. (Write the Society of Automotive Engineers, Inc., 400 Commonwealth Drive, Warrendale, Pennsylvania 15086.)

Barleycorn, Michael. *Moonshiners Manual.* Willits, California: Oliver Press, 1975.

Bradley, Jack. *Make Fuel in Your Back Yard.* Wenatchee, Washington: Biomass Resources. (Write to Biomass Resources, 625 Sage Hills Drive, Wenatchee, Washington 98801.)

Brown, Michael H. *Brown's Alcohol Motor Fuel Cookbook.* Cornville, Arizona: Desert Publications, 1979.

*Corn . . . Alcohol . . . Farm Fuel! Things You Need to Know.* Des Moines, Iowa: Iowa Corn Promotion Board, 1979.

Dabney, Joseph Earl. *Mountain Spirits.* Lakemont, Georgia: Copple House Books, 1974.

*Ethyl Alcohol For Fuel Use.* ATF-P 5000.1. Washington, D.C.: U.S. Department of the Treasury, Bureau of Alcohol, Tobacco, and Firearms. (Write the U.S. Government Printing Office, Washington, D.C. 20402.)

*Fact Sheet: Conversion of Automobiles to 100% Ethanol.* Washington, D.C.: National Center for Appropriate Technology. (Write the National Center for Appropriate Technology, 815 15th Street, N.W., Washington, D.C. 20005.)

*Feed Formulation.* Cincinnati, Ohio: Distiller's Feed Research Council. (Intended for the feed industry, but very helpful for the farmer who mixes his or her own feed. Write to the Distiller's Feed Research Council, 1435 Enquirer Building, Cincinnati, Ohio 45202.)

*Gasohol—Are Alcohol Fuels in Our Future?* Washington, D.C.: Citizens' Energy Project, 1979. (Write Citizens' Energy Project, 1110 6th St., N.W., Washington, D.C. 20001.)

*Gauging Manual.* ATF-P 5110.6. Washington, D.C.: U.S. Department of the Treasury, Bureau of Alcohol, Tobacco, and Firearms. (Instructions and tables for determining the quantity of distilled spirits by proof and weight. Write the U.S. Government Printing Office, Washington, D.C. 20402.)

Gibat, N. and K. *The Lore of Still Building.* Norman, Oklahoma: Popular Topics Press, 1973.

*Handbook of the Nutritional Contents of Foods.* Washington, D.C.: United States Department of Agriculture. (The largest and most detailed source of food nutrition information ever compiled, with complete content analyses of virtually every food—in all forms—available today. Write the Department of Agriculture, Washington, D.C. 20250.)

Lincoln, J.W. *Methanol and Other Ways Around the Gas Pump.* Charlotte, Vermont: Garden Way Publishing, 1976.

Mathewson, S.W. *Manual for the Home and Farm Production of Alcohol Fuels.* Los Banos, California: J.A. Diaz Publications, 1979. (Write J.A. Diaz Publications, P.O. Box 709, Los Banos, California 93635.)

Miller, Brinton M., and Warren Litsky. *Industrial Microbiology.* New York: McGraw-Hill Book Company, Inc., 1976. (Technical book on enzymes and fermentation.)

Ohm, Karl. "Gasohol Flows in Michigan." *Michigan Farmer*, Vol. 271, No. 5, pp. 4–9.

Pleath, S.J.W. *Alcohol, a Fuel for Internal Combustion Engines.* London: Chapman and Hall, 1949.

Prescott, Samuel Cate, and Cecil Gordon Dunn. *Industrial Microbiology.* New York: McGraw-Hill Book Company, Inc., 1949. (Technical book on the production of alcohol from various types of raw materials. Out of print.)

Scheller, William A. *Nebraska 2 Million Mile Gasohol Road Test Program—Sixth Progress Report.* Lincoln, Nebraska: University of Nebraska, Department of Chemical Engineering, January 31, 1977.

*Small-Scale Fuel Alcohol Production.* Washington, D.C.: United States Department of Agriculture, March 1980. (Write the Department of Agriculture, Washington, D.C. 20250.)

Solar Energy Information Data Bank. *Fuel From Farms—A Guide to Small-Scale Ethanol Production.* Washington, D.C.: Solar Energy Research Institute, Department of Energy, February 1980. (Write the Technical Information Center, U.S. Department of Energy, P.O. Box 62, Oak Ridge, Tennessee 37830.)

Stone, Charles L. *Synthetic Fuels Program.* Sacramento, California: California State Legislature, 1979. (Write the Assembly Publications Office, P.O. Box 90, State Capitol, Sacramento, California 95814.)

Tsao, G.T. *Production of Alcohol Fuel From Grain and From Cellulosic Materials.* July 23, 1979. (Testimony before the Agricultural Research and General Administration Subcommittee of the U.S. Senate Agriculture Committee.)

The University of Santa Clara and the University of Miami. *Comparative Automotive Engine Operation When Fueled With Ethanol and Methanol.* HCP/W1737-01. Washington, D.C.: U.S. Department of Energy, 1978. (Write the U.S. Government Printing Office, Washington, D.C. 20402.)

Van Winkle, Matthew. *Distillation.* New York: McGraw-Hill Book Company, Inc., 1967. (Engineering book on designing of fractional distillation stills.)

Wiebe, R. *Dual Carburetion With Alcohol-Water Mixtures and Alcohol Blends.* Peoria, Illinois: Northern Regional Research Laboratory, 1954.

Wiebe, R., and J.D. Hummell. "Practical Experiences With Alcohol-Water Injection in Trucks and Farm Tractors." *Agricultural Engineering,* Vol. XXXV, No. 5 (May 1951), pp. 319–326.

Willkie, Herman F., and Paul J. Kolachov. *Food for Thought.* Indianapolis, Indiana: Indiana Farm Bureau, Inc., 1942. (Experimental data on mashing of corn for alcohol.)

Also:
Gasohol U.S.A. (10008 E. 60th Terrace, Kansas City, Missouri 64133) is a good source of information relating to the production of gasohol fuel.

THE MOTHER EARTH NEWS® (P.O. Box 70, Hendersonville, North Carolina 28791) is a bi-monthly magazine which regularly features articles on alcohol fuel production and use.

# DEFINITION OF TERMS

**accelerator pump**—A single-stroke mechanical pump, activated by depressing the accelerator pedal, which injects a small stream of additional fuel down the throat of the carburetor to enrich the air/fuel ratio during acceleration.

**aeration**—Charging a liquid with a gas (or air) so that some of the gas is dissolved in the liquid. This is the free gas in the solution and is available to the organisms (such as fish or, in our case, yeast and bacteria) that live in the solution.

**aerobic**—Able to live, grow, or take place only where free oxygen is present. Mash can be aerated with oxygen by whipping it or by bubbling it with compressed air. This fixes a certain portion of free air in the liquid mash. Yeast needs this free oxygen or air to reproduce rapidly and build up a large colony. Oxygen or air in the mash provides an aerobic state for the yeast to live in.

**anaerobic**—Able to live and grow where there is no free air or oxygen. Yeast, along with many bacteria, is also able to thrive in an anaerobic state, an environment which lacks a significant amount of free air or oxygen. The anaerobes get oxygen from the decomposition of the compounds that they are acting upon.

**BATF or ATF**—Bureau of Alcohol, Tobacco, and Firearms.

**BDG**—Brewer's dried grains. Brewer's spent mash has a different grain composition from distiller's spent mash. It, too, is a good feed supplement.

**bottoms**—The residue left after distillation. However, in petroleum engineering the product taken off the bottom of storage tanks is also called the bottoms.

**compression ratio**—The amount of pressure applied to the air/fuel mixture within the combustion chamber during the compression stroke. It is determined by comparing the volume of the combustion chamber when the piston is at its lowest point (Bottom Dead Center) to the volume of the combustion chamber when the piston is at its highest point (Top Dead Center).

**DDG**—Distiller's dried grains. These are the solids strained from the spent mash and dried.

**DDGS**—Distiller's dried grains and solubles. The entire composition of the spent mash, once dried.

**downcomer**—Pipes which penetrate the plates of a still column, and may have caps or cups at their tops or bottoms. When water is separated from alcohol during distillation, the water falls toward the bottom of the still column through the downcomers. The water level rises on a plate until it reaches the height of the downcomer, then descends to the next plate.

**enzymes**—Protein biocatalysts produced by living cells. All living cells produce enzymes. When grain is malted (sprouted), certain enzymes already in the seed (such as beta-amylase) increase in number, and others (such as alpha-amylase) are newly formed. These enzymes cause a chemical reaction and degrade complex food into simpler food. However, they are very specific in what they hydrolyze. For example, the enzyme sucrase—which hydrolyzes sucrose—has no effect whatever on lactose or maltose, and so on.

**float valve**—A check valve within the carburetor which shuts off the incoming supply of fuel to the carburetor float bowl when the correct level within the bowl is attained.

**forged connecting rods**—High-strength connecting rods which are forged rather than cast in order to tolerate increased compression ratios without bending or twisting.

**forged crankshaft**—A high-strength crankshaft which is forged rather than cast in order to tolerate increased compression ratios without flexing.

**high-compression pistons**—Forged aluminum pistons with domed, notched heads which increase the compression ratio in an engine and still allow clearance for valve lift.

**hygroscopic**—Attracting or absorbing moisture from the air. Alcohol is hygroscopic. This is why it is impossible to keep 200-proof alcohol in storage very long (even if you could produce it), as it would soon dilute itself with available airborne moisture. A 170-proof alcohol fuel will not attract very much moisture from the air, so storage is really no problem.

**idle orifice**—The opening through which the air/fuel mixture passes during engine idle. The size of this orifice is adjusted through the use of a tapered idle mixture screw.

**intake manifold**—The assembly of channels which delivers the air/fuel mixture from the carburetor to each cylinder.
**main metering jet**—The fixed-size orifice (usually brass) which meters the correct amount of fuel from the float bowl to be mixed with air entering the air horn.
**mash**—The ground grain (or other products) and water composition that contains the sugars which have resulted from cooking and from the enzymatic conversion of starches.
**miscibility**—The ability to be mixed. When two or more liquids are totally miscible, they form a uniform blend . . . that is to say, they dissolve in each other.
**octane**—One component of gasoline which is used by the industry as a measurement of the fuel's ability to resist pre-ignition or engine knock. Octane is assigned the number 100 for fuel rating purposes, and—depending on the amount and type of antiknock compounds added to gasoline—the octane rating of the fuel can range anywhere from below 80 to above 100. The octane rating of super-premium automotive gasoline reached a peak of about 102 + in 1970, but today—depending on the grade of gasoline used—automotive ratings usually range from 87 to 92. There are four methods used to rate antiknock qualities, but for automotive purposes the only two in general use are the research method and the motor method, which differ only in the engine operating conditions and the means by which the measurements were taken.
**power valve**—A vacuum-controlled one-way valve within the carburetor which enriches the air/fuel ratio when the accelerator pedal is depressed.
**pre-ignition**—An undesirable situation that comes about when the air/fuel mixture is ignited prematurely because of either too-advanced spark timing, excessively lean mixture, or use of a low-grade fuel which cannot tolerate the intense compression built up in the engine's combustion chamber. Also referred to as "knock".
**proof**—A term used by the distilled spirits industry and the taxing agencies to give various types of alcohol products a common unit of measurement. The proof is just twice the percentage of the alcohol in the liquid. For example, an alcohol liquid that is 90 proof has only 45% alcohol in the total mixture. For taxing purposes, all alcohol is converted to *proof gallons*.
**proof gallon**—A standard U.S. gallon that contains 50% alcohol and 50% water (100 proof). An alcohol/water mixture that contains a different ratio of each can be translated into proof gallons. To do so, simply move the decimal point of the proof two places to the left and multiply by the total gallons of distillate. For example, an 85% alcohol liquid is 170 proof. 1.70 × number of gallons equals proof gallons.
**riser**—A tube which penetrates a plate in a column still and allows alcohol vapors to rise. There is always a cap or cup placed atop each riser to distribute the vapor into the column section and to prevent water from dripping into the riser.
**spark timing**—The critical point at which the spark plug fires in relation to the compression and power cycles in an internal combustion engine. The more advanced the spark timing is (short of pre-ignition), the more efficiently the engine runs and the more torque it will deliver.
**spent mash**—The solid and the liquid mash material left after fermentation and distillation are completed.
**stillage**—The same thing as spent mash. Often, no solids are removed from the mash composition before the distillation procedure . . . so the bottoms is stillage.
**supercharger**—A belt-driven air pump (sometimes called a blower) which greatly compresses the air/fuel mixture in the intake manifold and hence the combustion chamber, and thus effectively raises the engine's compression ratio.
**tax gallon**—The equivalent of a proof gallon but for beverage taxation purposes.
**thin stillage**—The liquid part of the spent mash with the solids removed.
**turbocharger**—An air pump, driven by the engine's escaping exhaust gases, which greatly compresses the fuel/air mixture in the intake manifold and hence the combustion chamber. The net effect is similar to that of a supercharger.
**wine gallon**—A volume measurement of alcohol or of an alcohol/water mixture. It is a standard U.S. gallon. The proof can be of any quantity. A 20% alcohol, 80% water (40-proof) composition that fills a gallon container is one wine gallon.
**wort**—The liquid portion of mash that has not yet been inoculated with yeast. A brewing term that defines the mash between the brewing-mash preparation and the fermentation period.

# SUBJECT INDEX